Food, Agri-Culture and Tourism

Katia Laura Sidali • Achim Spiller • Birgit Schulze
Editors

Food, Agri-Culture and Tourism

Linking local gastronomy and rural tourism: interdisciplinary perspectives

Springer

Editors
Ph.D. Katia Laura Sidali
Georg-August University of Göttingen
Department of Agricultural
Economics and Rural Development
Platz der Göttinger Sieben 5
37073 Göttingen
Germany
ksidali@gwdg.de

Prof. Achim Spiller
Georg-August University of Göttingen
Department of Agricultural
Economics and Rural Development
Platz der Göttinger Sieben 5
37073 Göttingen
Germany
a.spiller@agr.uni-goettingen.de

JProf. Dr. Birgit Schulze
Institute of Agricultural Economics
Christian-Albrechts-University of Kiel
Olshausenstr. 40
24118 Kiel
Germany
bschulze@ae.uni-kiel.de

Financed by

ATENEO ITALO-TEDESCO
DEUTSCH-ITALIENISCHES HOCHSCHULZENTRUM

DAAD Deutscher Akademischer Austausch Dienst
German Academic Exchange Service

ISBN 978-3-642-42229-4 ISBN 978-3-642-11361-1 (eBook)
DOI 10.1007/978-3-642-11361-1
Springer Heidelberg Dordrecht London New York

Cover design: eStudio Calamar S.L.

Printed on acid-free paper

Springer is part of Springer Science+Business Media (www.springer.com)

Contents

List of authors

Ludwig Arens is a graduate in business administration, and has been working for the Chair of Agri-Business Administration (Prof. Dr. Ludwig Theuvsen) since mid-2009. His research focus is the improvement of the quality of communication in agri-business. As a result of his family and his roots in the wine region Rheinhessen, he is strongly connected to the wine business. Although this article is strictly outside his research remit at the Georg-August University in Göttingen, he had a personal interest in contributing to its creation.

Rossella Bianchi studied Political Science at the University of Bari, and is now a researcher in the Department of Engineering and Management of Agricultural, Livestock and Forestry Systems of the Faculty of Agricultural Science. She has attended courses at the Centre for Advanced Economic and Agricultural Studies for Southern Italy at Portici (Naples), has been a visiting professor at the Institut für Ländliche Strukturforschung of the J.W. Goethe Universität, Frankfurt am Main, and a researcher at the University of Basilicata. Her research and teaching concern programming of agricultural economic development, environmental and institutional policies for marginal and mountainous areas in Southern Italy.

Stefano Buiatti has been working at the Department of Food Science of University of Udine since 1990. He gives courses on “Brewing Technology” (since 1994), “Cleaning and Disinfection of Food Plants” (since 1999) and “History and Culture of Food” (since 2007). He has carried out research at the Brewing Research International, Nutfield (UK) in 1997, 1998 and 1999. He has worked as an evaluator on the Scientific Advisory Board for EU projects since 1999.

Maurizio Canavari is Associate Professor of Agricultural Economics and Appraisal at the Alma Mater Studiorum University of Bologna. He holds a Laurea degree in Agricultural Sciences and a Doctorate degree in Appraisal and Land Economics. He teaches "Marketing management in agribusiness" and "Marketing research and planning" at the Faculty of Agriculture in Bologna. His research activity has covered a wide range of topics in agricultural economics. Recent research projects deal with trust and quality assurance in food networks, marketing of quality, organic, and unique food in European and Asian context, certification of organic products, consumer preferences towards genetically modified food. He has authored or co-authored over 80 conference/seminar papers, journal articles, book chapters and research reports, both at national and international level.

Alessio Cavicchi is a Researcher in Agricultural Economics. He received his PhD in Economics of Food and Environmental Resources from the University of Naples “Parthenope” and a Master of Science (MSc) in Food Economics and Marketing from the University of Reading (UK). His works have been published in several journals, including Food Quality and Preference, British Food Journal, Journal of Agricultural Economics, Agribusiness: an International Journal, International Journal of Wine Business Research and Food Economics, among others. Previously, he worked in Finance and had a Internship at the European Parliament in Brussels. His main fields of interest in research are consumer food choice, economics of food quality and safety, and innovation in the agro-food sector.

Axel Dreyer is Professor for Tourism Marketing at the Hochschule Harz, University of Applied Sciences, Wernigerode (since 1993) and Honored Professor for Sport Management at the University of Goettingen (since 1996), where he did his PhD. Axel Dreyer is co-founder of the German Society of Tourism Sciences (DGT - Deutsche Gesellschaft für Tourismuswissenschaft e.V.). At the competence center of Hochschule Harz he manages the Department of Tourism together with his colleague Sven Groß. Axel Dreyer has published a lot in the fields of tourism markets and tourism marketing. Special issues among others are service quality and customer satisfaction, the impact of demoraphic change on tourism, sports tourism and cultural tourism. At present he is working on hiking tourism, cycling tourism and wine tourism.

Matthias Fuchs is Professor at the European Tourism Research Institute (ETOUR), Mid-Sweden University, Sweden. He is also the scientific director at the eTourism Competence Center Austria (ECCA). His research interests include destination monitoring, tourism marketing, e-business adoption and impact, e-commerce as well as knowledge retrieval and management in tourism.

Amos Gyau was born in Berekum, Ghana. He obtained his bachelor's degree from the University of Ghana in Economics (Major) and Statistics (Minor) in 1998 and his master's degree from the University of Copenhagen in Agricultural/Development Economics in 2003. He furthered his education at the University of Göttingen in Germany where he obtained his PhD degree in 2008 specialising in supply chain management in international food and agricultural business. Currently, he works as a lecturer at School of Agriculture, Food and Wine at the University of Adelaide. His current research includes agricultural and food value chain analysis, consumer studies and wine and food tourism. He has published articles in international peer review journals and international conferences in these subject areas.

Wolfram Höpken is Professor for Business Informatics and eBusiness at the University of Applied Sciences Ravensburg-Weingarten. He is director of the E-Commerce Competence Centre Swabia (KECoS) and scientific director of the eTourism Competence Center Austria (ECCA). His main fields of interest are intelligent trip planning and recommendation systems, semantic web and interoperability, mobile services and business intelligence and data mining.

Rossana Mari is responsible for the project "Educational farms" in Emilia-Romagna region, she has been working for the local government since 1992. She holds a Laurea degree in Agricultural Science and in Literature and Arts at the University of Bologna. Her main activities include the following areas: communication on agriculture and quality food production, educational projects geared towards school children and consumers with the aim of promoting a healthy lifestyle, organisation of events to inform the general public about sustainable agriculture, traditional and typical food products. She has contributed to regional and national projects and programs. She collaborates with universities, public and private associations; she is the author and co-author of a number of articles and book chapters.

Juliane Müller completed her Masters degree in Tourism and Destination Development at the Hochschule Harz, University of Applied Sciences, Wernigerode, in 2009. She took part in several research projects in the field of wine tourism. In her Masters thesis she analysed suitable marketing instruments for wine and culinary tourism products. After that she published a book about wine tourism together with Prof. Dreyer. Now she is working as Assistant Product Manager Southern Africa for FTI Frosch Touristik GmbH, a large German tour operator.

Domenico Regazzi got his Laurea degree in Agricultural Sciences at the University of Bologna. Since 1972 he has been working at the Alma-Mater Studiorum – University of Bologna. In particular, since 1991 he is Full Professor in "Market Economies and Marketing". From November 1, 1999 to October 31, 2005 he was Dean of the Faculty of Agriculture. He is involved in research activities concerning the economic analysis of farms, economics of the production and distribution of fishing and agricultural products, agri-food marketing, and rural appraisal. He has been responsible for national and international research programs. He is author of about 200 scientific publications. Since 2004 he is chairman of the Italian Society of Agrifood Economics.

Francesca Regoli BA in Political Science, she is PhD candidate in International Cooperation and Sustainable Development Policies at the University Alma Mater Studiorum of Bologna. She has worked several years (2002-2007) for the Executive Secretariat of UniAdrion Network for the coordination and management of different projects of cooperation related to the Adriatic-Ionian Basin.

Cristina Santini is a Researcher at Unitel. She received her PhD in Economics and Management of Enterprises and Local Systems from the University of Florence. She received a Masters Degree in Marketing and Management for Wine Business at the University of Florence. She served as lecturer on Management and International Management at the University of Florence in Italy. Previously, she worked as marketing manager in a Tuscan winery and has had some work experience in a multinational company. She currently works in research and is a consultant with the Italian Centre for Sensory Analysis in Italy. Her research interests are Entrepreneurship, Innovation and Small Business.

Andreas Scharf is Professor for Business Administration, specialising in Marketing and Marketing Research, at the University of Applied Sciences of Nordhausen (Germany). He also works as Scientific Head at ISI, Institute for Sensory Analysis and Innovation Consulting of Göttingen. He has published in both national and international journals and some of his contributions are important milestones in the literature of sensory analysis.

Kai and Jan Schätzel The Schätzel family has been growing wine in the Rhine region for 650 years. The family business developed from mixed enterprise (farming and viticulture) to solely marketing bottled wine. The vinery was initially based in Guntersblum. The second important strand in the story of the Schätzel vinery is the Schlamp family and their vinery in Nierstein. The Schlamp vinery once counted among the most prestigious businesses in the winegrowing region of Nierstein on the Rheine, whose main centre was, and remains, a farmhouse on the Oberdorf Street in Nierstein. In 1996, the future heir Kai Schätzel began to take an active role in the business. At that point just leaving school, Kai initially looked after the cultivation of the wine before increasingly becoming involved in the business. During this time, he gained experience on premium wineries in California, and in 2009, ownership of the Schätzel vinery was transferred to him. Jan Schätzel is the brother of the current owner Kai. In 2009 he passed his first State Law exam in Göttingen. He too because of his heritage has a strong interest in viticulture business structures, and for this reason has contributed to this article.

Birgit Schulze is Junior Professor for Agribusiness and Supply Chain Management at the University of Kiel since 2010. She took her PhD in 2007 at the University of Göttingen. Her main research interest is the analysis of agri-food supply chains with respect to chain actors' behavior, i.e., from consumers to farmers. A special focus has been on supplier relationship management in milk and meat chains. Her work has appeared in the Journal on Chain and Network Science, Züchtungskunde, Journal of the Canadian Association for Leisure Studies Leisure/Loisir and a number of national and international conference proceedings.

Andrea Segrè PhD, is Dean of the Faculty of Agriculture of the University Alma Mater Studiorum of Bologna, Professor of Agricultural Policy and Coordinator of the PhD in International Cooperation and Sustainable Development Policies. He has dealt with the transition processes of the economical-agricultural systems in the countries of the real socialism, developing many missions (from 1990 to 2000) on the field on behalf of various International Organizations (European Commission, World Bank, FAO, OCSE, Italian Ministry of Foreign Affairs) and publishing different volumes and scientific articles. He has designed, implemented and promoted the Last Minute Market projects, to reclaim food and non food unsold products and distribute them to charity associations (www.lastminutemarket.org).

Katia Laura Sidali is post-doc researcher at the University of Göttingen (Germany) since 2009. She took her PhD in a joint doctoral program between the University Alma Mater Studiorum of Bologna (Italy) and the University of Göttingen. Her research focuses on marketing of food and tourism, consumer behavior and communication strategies in tourism. She has been awarded scholarships by the Italian Trade Commission (ICE) in 1993 and by the German Steinbeis Association in 2000. Her research appears in the following reviewed scientific journals: Journal of Global Marketing Science, Journal of the Swiss Society of Agricultural Economics and Agricultural Sociology and Journal of the Canadian Association for Leisure Studies Leisure/Loisir.

Roberta Spadoni is Tenured Assistant Professor at the Department of Agricultural Economics and Engineering, Alma Mater Studiorum University of Bologna. She holds the Laurea degree in Agricultural Science at the University of Bologna and Doctorate degree in "Economics of the agri-food systems" at the University of Parma. She teaches "Agricultural and industrial marketing" at the Faculty of Agriculture in Bologna. Her research activity has covered the following areas: economics of agricultural and food markets, certification systems, agricultural and industrial marketing, product quality issues. Recent research projects deal with Corporate Social Responsibility and quality standards in the food systems. She has authored or co-authored over 80 conference/seminar papers, journal articles and book chapters, both at national and international level.

Achim Spiller is Professor and Chair of Marketing of Agricultural Products and Food, University of Göttingen, Germany, Department of Agricultural Economics and Rural Development since 2000. His research interests centre around quality management, consumer behavior, branding and organic food marketing. He is a member of the Scientific Advisory Group of the Federal German Ministry of Agriculture and serves as Chair of the Stakeholder Board of the German Quality Certification System QS GmbH. His work has been published in both national and international journals (more details are available at: http://www.uni-goettingen.de/en/20076.html).

Nina Stockebrand has been a research assistant at the Chair Marketing of Food and Agricultural Products at the University of Göttingen since 2007. Her research topics include organic food, organic retail and specialized food retail as well as marketing of regional and organic food products. As a student of agricultural science she was involved with studies on organic food and sustainability communication in food retail. She gained practical experience in quality management at the German retailing company Tegut.

Randy Stringer is a professor in agricultural and food policy and the head of the Global Food and Agricultural Business at the University of Adelaide. Over the past 30 years, Prof. Stringer taught, published and conducted research and policy analysis on agricultural development, natural resource management, food security, land tenure and environmental issues in Asia, Africa, Europe and Latin America. Before joining the University of Adelaide, Prof. Stringer was the chief of the Comparative Studies Service at the United Nations Food and Agricultural Organisation.

Claudio Visentin is Professor for "Cultural History of Tourism" and "Heritage Interpretation" at the University of Lugano. At first he devoted himself to the study of International Relations, with special regard to intercultural dialogue, and he has published a book about the image of Germany in Italy ("Nel paese delle selve e delle idee. I viaggiatori italiani in Germania 1866-1915", Jaca Book ed., Milano 1995). He then turned to the history and culture of travel and tourism, and he has particularly investigated the history of Italian tourism and package holidays ("Il turismo contemporaneo. Cultura e mondo dell'impresa", Giampiero Casagrande ed., Lugano 2002). He researched the image of Switzerland as portrayed in international guidebooks ("Il Canton Ticino visto dagli altri. L'immagine del territorio nelle guide turistiche internazionali," Giampiero Casagrande ed., Lugano 2007). He is General Secretary of the International Commission for the History of Travel and Tourism, affiliated to the International Committee of Historical Sciences. He is also a member of the Editorial Board of the scientific reviewed Journal of Tourism History (more details are available at: www.claudiovisentin.it).

Matteo Vittuari PhD in International Cooperation and Sustainable Development Policies at the University Alma Mater Studiorum of Bologna, has been postdoctoral fellow at the Center for International Development at Harvard University. He has been visitor researcher in a number of universities and research institutions, including FAPRI – Food and Agricultural Research Institute – University of Missoury, the K.U. Leuven, the University of Banja Luka.

Preface

Katia L. Sidali[a]
Birgit Schulze[b]
Achim Spiller[a]

[a]University of Göttingen
Department of Agricultural Economics and Rural Development
{Katia-Laura.Sidali, a.spiller}@agr.uni-goettingen.de

[b]University of Kiel
Institute of Agricultural Economics
bschulze@ae.uni-kiel.de

Originating from the need stressed by several scholars to devote more research to the complex relationship between the local food producers and processors on the one hand and the tourist industry on the other, this book aims to build a bridge between these two realities.

Having practised fruitful research cooperation on farm tourism between the Universities of Göttingen and Bologna for several years, the editors of this book aim to broaden the scope of both the involved universities and the research streams. To this purpose, in December 2009, a conference on "Food, Agri-Culture and Tourism" took place in Göttingen under the aegis of the DIH (Deutsch-Italienische Hochschulzentrum) and the DAAD (Deutscher Akademischer Austausch Dienst). This book comprises the work presented at this event as well as combining further results and best practice case studies from scholars and practitioners from more than eleven universities.

To this purpose, the book has an interdisciplinary approach. Regarding the structure, there are three main sections. The first part is entitled "Farm and rural tourism" and analyzes the tourist sector in many European countries. Stockebrand, Sidali and Spiller compare different communication styles to promote high-quality food in German farm tourism. Regoli, Vittuari and Segrè focus their analysis on the recent evolution of rural and farm tourism in Romania. Finally, Bianchi offers a diachronical overview of agritourism in the context of the common agricultural policy and comments on the Italian law No. 96/2006 which is entirely dedicated to the promotion of agritourism.

As the title already suggests, the second part of the book links the topic of farm and rural tourism with the sector of food and wine. As shown by Canavari et al. farm operators who tailor educational activities for school groups can make an important contribution to improve the nutritional habits of these young farm guests. Dreyer and Müller claim that cooperative marketing can display economic benefits also at a regional level, using the example of the synergies established between tourist operators and wine makers in the region of Saale-Unstrut. Thus, the challenges posed by the rapidly changing behaviour of customers seem to be well managed by farmers who are eager to try new ways to attract guests and appreciate acting together. Creativity is another important driver of success, as it is shown in the best-practice described by Schätzel, Arens and Schätzel. Buiatti stresses the important role of other intermediaries, such as the Slow Food Association, who mediate between individuals and farmers. The latter in fact often ignore the new sensitivity of customers for a more sustainable and authentic way of food production.

Other strategies for helping farmers to better profile their oeno-gastronomical offer as a unique selling proposition are discussed in section three which is entitled "New avenues of research: online marketing and sensory analysis". While Gyau and Stringer as well as Fuchs and Höpken show the pros and cons of adopting e-marketing strategies in the farm and rural hospitality industry, Santini, Cavicchi and Canavari as well as Scharf focus on the growing importance of sensory analysis for better identification of market-segments based on their sensory attributes.

Hence, as stressed by Telfer and Wall in 1996 "relationships between tourism and food production can be placed on a continuum ranging from conflict through coexistence to symbiosis". This book claims, more than 10 years later, that above all rural and farm tourism has proven to be an example of such a symbiotic relationship between tourism and the food industry.

The Editors

Food, Agri-Culture, and Tourism

Claudio Visentin

USI The University of Lugano
Faculty of Communication Sciences
claudio.visentin@usi.ch

Interdisciplinarity is not the mere merging of different disciplines – each called into account to shed light from its specific perspective onto the subject of study. Rather, interdisciplinarity is a means to identify and study new themes that single disciplines would not be able to put into focus, describe and interpret by themselves. This is precisely what has been achieved during the past few years by the cooperation between Göttingen and Bologna that has then been extended to include universities from other countries both in Europe (such as Austria, Sweden and Switzerland) and beyond (e.g., Australia). Such a remarkable mobilisation of competences and methodologies has allowed us to consider the interrelationship between **food**, **agriculture** and **tourism** from a novel perspective. Food, agriculture and tourism are three variables, three spheres undergoing rapid and simultaneous transformation – generating links and combining themselves in innovative forms.

Food and wine change. Anything concerning food and wine culture is usually met with great interest in the public and media. Indeed, wine and food production and consumption are widely acknowledged global phenomena, dealt with by multinationals. Nevertheless, there are also many aspects of innovation, opening up spaces for new models, e.g., typical products, farmers' markets, or the consumption of locally sourced products. From our perspective, a significant trend is the one linking the consumption of local products to their place of origin – the idea that a specific food or wine can be fully enjoyed and "understood" only if it is prepared in its territory of origin, by people from that community, with local ingredients. Also the tasting of food and wine tend to take place increasingly in the same locus of production, providing at the same time an occasion to deepen the knowledge of the techniques used for the production and preparation (i.e., the material culture) of which that specific wine or food is a part of. In this way, oeno-gastronomy, food and wine tourism, and cultural tourism combine themselves in novel forms, which are difficult to dissect. Nowadays these acknowledgments may seem to be a foregone conclusion, but it is worth remembering that this was not at all the case until just a few years ago.

Agriculture changes. This primary sector is still facing structural problems: the countryside desertion by the younger generations; scarce profitability and the subsequent dependence on sources of external funding such as the EU, etc. In the light of this, the agricultural sector is taking a greater interest in integrative activities such as tourism that could help improve its condition and redefine its social role – creating, for example, new links and relationships with younger generations and schools, via programmes of food and environment related education.

Tourism changes. The new tendencies of contemporary tourism, timely followed by media attention, seem to point towards a predilection for independent trips, easily accessible holiday destinations (not too far from home), slow rhythms, a propensity to explore traditions and a strong accent on the authenticity of the experience and environmental sustainability. These lifestyles are uncommon in everyday life, but they can be re-discovered and experimented during the "altered" and almost "suspended" time of holiday. Such shifts in tourism demand have sustained the recent growth of farm holidays (*agritourism*) throughout Europe. Moreover, they have aided the increase in tourism in regions blessed with beautiful landscape and/or rich cultural heritage, providing a vital source of income for the local businesses quick to take advantage of these opportunities. In this field, Italy represents a model by which many other countries (beginning with Germany) are looking at with great interest. Taking precedence over any other country, Italian agritourism has now achieved full development – showing a tendency to be structured around very diverse models, referring to two wide areas. In some places, tourism is seamlessly integrated into the main agricultural activity, as in the case of Alto Adige. Guests "live" on the farm, consume its products and sometimes take part in its activities. In other Italian regions (such as Tuscany, Umbria or Apulia) tourism has clearly superseded agricultural activities. Such a choice could obviously be considered opportunistic, but also mirrors the diverse level of involvement that tourists are willing to accept. If the tourist interest in the agricultural world is no more than a mild curiosity, with no real desire to get directly involved in the everyday life of the countryside, it is logical to propose a more "sheltered" and "diluted" experience. This would often offer a high level of comfort (the swimming pool!), but is essentially distant from the daily life of a farm. Such a model would work well for agritourism located near "art cities" (Florence, Siena, Perugia, etc.), as they become a base for traditional cultural tourism. Another example is that of ancient, deserted villages, which are restored, transformed and proposed to tourists. In these cases, the absence of any concrete link with the world of agriculture and the farming community instils doubts about the appropriateness of the term "agritourism", and seems to suggest the need to develop a novel terminology.

The interconnected transformation of all these spheres has happened at a time when new technology has started a dramatic acceleration in the types of touristic communication. Although studies that focus on how the revolution of the Internet has changed the contemporary world are commonplace, there is little reflection on how tourism is one of the sectors that has been most influenced by these innovations. A new customer-led approach, combined with the tourist's capacity to get in touch directly with the touristic offer (bypassing traditional forms of intermediation), definitely favours farm holidays. Furthermore, small producers have also learnt how to circumvent their traditionally isolated position via their own web pages – establishing, at a paltry cost, a network of contacts with potential clients worldwide.

The meeting point of all these concurrent and contemporary transformations opens up many new opportunities for the agricultural world. There are certainly numerous possibilities – but some may be more attainable than others. The future of the agricultural world could well lie in the (difficult) search for a meeting point between continuing to make products for the commercial market and direct sale of high-quality food products (e.g., organic, geographically protected), forms of hospitality for tourists, educational projects for schools, etc. New forms of modern communication (spanning from digital to sensorial), could sustain all these ambitions. It is evident that many farmers will find it difficult to meet such complex demands, especially if they are not supported by guidelines drawn from both strategic reflection and original research, which are then passed on through appropriate training programmes.

The Italo-German cooperation developed over the years, however, has certainly proven able to gain a good start on this path.

to circumvent their traditionally isolated position was their own web pages establishing, at a paltry cost, a network of contacts with potential clients worldwide.

Part I

Farm and rural tourism

A sideways look at farm tourism in Germany and in Italy

Katia L. Sidali

Georg-August University of Göttingen
Department of Agricultural Economics and Rural Development
Katia-Laura.Sidali @agr.uni-goettingen.de

Abstract

The aim of the present contribution is to investigate farm tourism, using a comparative international approach (Gerrmany and Italy) and, thereby, to introduce important topics of the book. These two countries were chosen because of the complementary nature of their markets for agri-tourism. The image of agritourism in Italy, for example, is characterized by high quality and exclusiveness because it is inseparable from the gastronomic choices in food and wine. The sector also has weaknesses which lead, among other things, to a delayed development in strategic organisational planning. In addition in Germany, farm tourism has, up to now, only caught on with a limited part of the population. In contrast to Italy, its image is, however, somewhat modest. This form of tourism originated in a social tradition which has made agritourism particularly interesting for, above all, families with children. A difference from the Italian market lies in the fact that the German farm holiday has been successful in establishing a well-developed system of quality certification.

Zusammenfassung

Ziel des vorliegenden Beitrags ist es, den Wirtschaftszweig Urlaub auf dem Bauernhof mit einer international vergleichenden Betrachtungsweise (Deutschland und Italien) zu untersuchen und somit in wichtige Themen des Buches einzuführen. Die Auswahl dieser beiden Länder beruht dabei auf der Komplementarität der jeweiligen Märkte für Agritourismus. Beispielsweise besitzt der Agritourismus in Italien ein Image, das durch Hochwertigkeit und Exklusivität gekennzeichnet ist, nicht zuletzt deshalb, weil es untrennbar mit dem gastronomischen Angebot an Weinen und Speisen verbunden ist. Der Sektor weist allerdings auch Schwächen auf, die unter anderem auf eine verspätete Entwicklung von strategischen Organisationsplänen zurückzuführen ist. Auch in Deutschland erreicht Urlaub auf dem Bauernhof bislang nur einen begrenzten Teil der Bevölkerung. Im Gegensatz zu Italien ist sein Image jedoch eher bescheiden. Diese Form des Tourismus entstammt einer sozialen Tradition, die den Agritourismus insbesondere für Familien mit Kindern interessant macht. Im Unterschied zum italienischen Markt ist es dem deutschen Urlaub auf dem Bauernhof jedoch gelungen, ein gut entwickeltes System zur Zertifizierung von Qualität zu schaffen.

Riassunto

In questo articolo il mercato agrituristico viene analizzato con un taglio internazionale (Germania-Italia) al fine di introdurre alcuni temi ricorrenti in questo libro. Questi due Paesi sono stati scelti per la complementarietà del mercato agrituristico che presentano. In Italia per

K.L. Sidali et al. (eds.), *Food, Agri-Culture and Tourism: Linking local gastronomy and rural tourism: interdisciplinary perspectives*, DOI 10.1007/978-3-642-11361-1_1,

esempio l'agriturismo gode di un'immagine raffinata. Infatti, essendo indissolubilmente legato all'offerta enogastronomica, il target di mercato è costituito dalla fascia medio-alta della popolazione. Pur partendo da queste premesse, l'agriturismo italiano rimane un mercato di nicchia anche perché l'elaborazione di piani di organizzazione strategica da parte degli agenti del settore è abbastanza recente. In Germania il settore agrituristico riguarda solo un' esigua fascia della popolazione, perciò la sua immagine è più modesta. Questa forma di turismo, infatti, deriva dalla tradizione popolare che lo rende interessante soprattutto per famiglie con bambini. A differenza del mercato italiano, tuttavia, l'agriturismo tedesco è riuscito a creare un sistema di certificazione di qualità molto avanzato.

Keywords: Farm tourism; Rural tourism; Cross-country comparison.

1 State of the art of German farm tourism

1.1 Definition and typology

Defining *Urlaub auf dem Bauernhof* (German name for farm tourism) is problematic. Firstly, in the literature there are many different definitions of this form of tourism (Busby and Rendle, 2000; Roberts and Hall, 2001). Secondly, in Germany comprehensive statistics on farm tourism are lacking because the majority of farm operators, who are very small and privately managed (Oppermann, 1996, p. 93), are automatically included in the private operators hospitality directories (*privat Vermieter*) (BMELV, 2008). As a consequence, the majority of studies rely on estimations. This situation is compounded by the fact that Germany does not have legislation specifically dedicated to farm tourism. Thus, farm operators are required to abide by several rules under civil law, as in the case of construction of new buildings (*Baugesetzbuch*), rural area regulations (*Landes-Gaststättenbau-Verordnung*), agrarian law, etc.

Whenever farm tourism is offered by a farm operator who hosts up to eight guests, the business is considered agricultural property management (*Vermietung als Vermögensverwaltung*), and no official registration is necessary. Above the eight-bed threshold, a holiday farm is subject to public catering law (*Gaststättengesetz*). The latter involves additional work (Oppermann, 1996, p. 93), which explains why the majority of farmers stay below this limit. Furthermore, in Germany farm tourism can also be offered on farms where there is no longer any farming activity. As a consequence, it is particularly difficult to demarcate this sector from other rural tourism operators, such as B&Bs and country hotels.

Henceforth, taking the legal framing as a point of departure for defining German farm tourism we will use the definition given by Przezbórska (2003), who explains farm tourism as "all tourism and recreation activities connected with a [...] farm or any agricultural, horticultural, fishery or agribusiness operation."

According to Barth and Theis (1998), we can classify farm operations based on their location, dividing them into mountain, hill, seaside or natural park farm operations. Another way to classify them is based on the grade of intensity of the farm experience that farm operators provide to their guests. A high-intensity farm experience implies a wide variety of farming-related products or activities (for example, the opportunity for guests to work on the farm). In contrast, some farm operators provide only accommodation. Another possibility to classify farm tourism operations is by the type of accommodation offered to farm guests. Farm vacationers can obtain lodging either in independent accommodation, such as apartments, holiday houses, etc., or in rooms. Furthermore, farm guest lodging in tents or caravans is gathering momentum (AgE, 2007a).

Quality certification is also a reliable means of categorizing farm operations, which can be classified according to national systems (like the hotel star system, the DLG system and the BAG system) or regional ones (such as the flowers classification in Bavaria). Finally, farm tourism operations can be classified according to the financial contribution of agricultural activities in relation to the tourist ones: if the former predominate, agritourism is only a side activity otherwise it is the main activity. Table 1 gives an overview of various defining criteria applied to farm operations.

Table 1. Criteria for the classification of farm tourism

Defining criteria	Example
Legally set threshold	Agricultural property management Public catering law
Location	Mountain Sea Natural parks Lakes
Type of farming experience	High-intensity farming experience Low-intensity farming experience
Type of lodging	Apartments Rooms House Tent Caravan
Classification system	National (star system, DLG certification, etc.) Regional (flowers in Bavaria, etc.)
Financial contribution of agritourism activities	Agritourism as main activity Agritourism as side activity

Source: own representation based on Barth and Theis (1998, p.14)

1.2 Historical development

The origin of farm tourism in Germany dates back 150 years (Nilsson, 2002), but it was after the Second World War that this sector began to gather momentum (Wagner, Burger and Magnus, 1997). Farm tourism boomed in the 1970s as is shown by the large number of studies of that time (Klöpper, 1974; Lehle, 1982; Mook, 1978; Schulz-Borck and Tiede, 1974; Tiede and Schulz-Borck, 1975). It was, however, during the 1990s that this tourist sector really gained national visibility. During this period uncertainty in the agricultural market was counteracted by significant investment in farm tourism, turning it into one of the common goals of European regional policy (Skuras, Petrou and Clark, 2006). As a consequence, farmers began to renovate their buildings in order to turn them into appealing tourist accommodations (Caballé, 1999).

According to the German Ministry for Food, Agriculture and Consumers' Protection (BMELV, 2006), German farm tourism reached its peak in late 1996 with 3.9 million guests but dropped again in the period 1999-2001 before gradually recovering. However, due to cyclic fluctuations, the trend for this form of tourism is particularly difficult to forecast (BMELV, 2008).

1.3 The supply side

In Germany, farm tourism is currently offered by around 25,000 farm operators and generates a yearly turnover of € 943 million (BMELV, 2006). The states of Bavaria and Baden-Württemberg are the most important destinations, followed by Schleswig-Holstein and Lower Saxony (ibid.).

It is estimated that, for the majority of German farm operations (BWT, 2008; Lemke, 2003; Sidali, Schulze and Spiller, 2007), farm tourism activities contribute between 25% and 35% of overall income (see Table 2). According to several studies (Lemke, 2003; Sidali et al., 2007; Wagner et al., 1997) the mean age of German farm tourism operators is around 50 years old. Furthermore, German farm tourism involves numerous female farm operators, which is in line with the international trend.

Table 2. Supply side of German farm tourism

Supply side of German farm tourism	
Market share of agritourism in Germany	0.6%[1]
Total number of farm tourism operations	25,000
Total revenue (millions €) (2006)	943
Share farm tourism revenue : total revenue	25%-35%
Participation of female farm operators	High
Mean age of farm operators (years)	50
Occupancy rate (nights per year)[2]	Between 125 - 135

Source: BAG, 2008; BMELV, 2006; BMELV, 2008; Lemke 2003; Sidali et al., 2007

The main associations responsible for promoting this tourist sector are both agriculture-based organizations, such as the DLG (Deutsche Landwirtschaft Gesellschaft) and BAG (Bundesarbeitsgemeinschaft für Urlaub auf dem Bauernhof), and tourist-based organizations, such as the DZT (Deutsche Zentrale für Tourismus) and the DTV (Deutscher Tourismus Verband).

According to the BMELV (2008), the main competition facing German farm operators comes from the following sectors: campsites, hotels and foreign holiday

[1] Calculated as follows: ratio of the total revenue of the German agritourism (€ 943,000,000) (BMELV, 2006) to the total revenue of the German tourist sector (€ 150,000,000,000) according to the DTV (German Tourism Board) ([www.deutschertourismusverband.de /content/files/zdf% 202007.pdf] November 2008)

[2] Total number of nights per year during which beds are occupied

destinations. With regard to the latter, it is interesting to point out that many Germans state that they are not interested in farm tourism in their own country; however, they are interested in this tourism form abroad (especially in Austria, Italy and Spain) (see Table 3).

Table 3. Competitors

Competitors	
Market share of hotels in Germany	12.3%[3]
Market share of camping sector in Germany	0.8%[4]
Market share of agritourism in relation to hotels	5.0%
Market share of agritourism in relation to camping	79.9%

Source: BMELV, 2006; BMELV, 2008; Lemke, 2003; Sidali et al., 2007

1.4 Market demand and potential

According to the survey carried out on behalf of the German government (BMELV, 2006), the typical farm guest is 44 years old with an intermediate level of education and monthly income. Large-sized families dominate this guest group. The main motivations for having a farm holiday are to escape day-to-day life, to relax and to recuperate. For these reasons, the main activities on the farm are excursions, sampling regional products and relaxing (Table 4).

Several studies estimate that farm tourism is still not achieving its real market potential. According to the German government (BMELV, 2006), for instance, in the period from 2006 to 2008, around 2.5% of Germans indicated that they were definitely planning a farm holiday, and 8% said that they were probably going to do so. In contrast, among international guests the demand for German farm tourism is almost nonexistent.

[3] Calculated as follows: ratio of the total revenue of the German hotel sector (€ 18,500,000,000) according to DEHOGA (German hotel federation) ([www.dehoga-bundesverband.de/home/branchenthemen_ 0_1009.html]November 2008) to the total revenue of the German tourist sector (€ 150,000,000,000)

[4] Calculated as follows: ratio of the total revenue of the German camping sector (€ 1,180,000,000) according to the BMELV (2008) to the total revenue of the German tourist sector (€ 150,000,000,000)

Table 4. Demand side of German farm tourism

Demand side of German farm tourism	
Total number of farm guests 2003-2005 (millions guests)	2.10
Average German holiday farm guest	
Mean age (years)	44
Monthly income (€)	2,000
Education	Intermediate
Motivations for choosing a farm holiday	**Activities during a farm holiday**
1. Escape day-to-day life	1. Excursions
2. Relaxation	2. Sampling regional specialties
3. Recuperation	3. Relaxation

Source: BMELV, 2006; BMELV, 2008

1.5 Strengths and weaknesses of German farm tourism

The aim of this section is to provide an overview of the agritourism sector in Germany based on an analysis of the key literature from the 1970s to present. It is also based on a variation of the SWOT (Strengths, Weaknesses, Opportunities and Threats) model suggested by Johnson and Scholes (1997).

Strengths

According to Bodenstein and Spiller (1998) strengths encompass everything a business does better than its competitors and all the positive product features within the operator's control. Hence, these are the characteristics operators should build on. According to the literature, the main strengths of German farm tourism are intangible assets, which constitute a clear competitive advantage over other tourist sectors.

"As this form of tourism is provided by the locals, and the benefits mainly remain in the region" (Embacher, 1994), farm tourism has been recognized as a form of sustainable development for less developed areas. Furthermore, its environmentally friendly nature has led to farmers being recognized as guardians of the landscape, or "environmental architects" (Roberts and Hall, 2001). This has been picked up by the government (Nilsson, 2002), which has, as a consequence, recently renewed its support for the "further development" of farm and rural tourism in order to economically support rural areas (German government press release, September 24, 2008). Furthermore, during the last few decades, a large number of programs aimed at enhancing farm tourism have been commissioned by the European Union (above all, the LEADER program) (Nilsson, 2002; Skuras et al., 2006).

Farm tourism operators are characterized by a particularly high commitment (Sidali et al., 2007), which depends on several factors. First of all, tourism allows many farmers "to remain self employed and work solely on the farm" (Embacher, 1994), which is "a key element of the farmers' identity" (ibid.). Furthermore, other reasons, including companionship with guests and the (farmer's) "wife's convenience" (Nilsson, 2002), motivate farmers to diversify their activities and enter into this business (McGehee and Kim, 2004; Nickerson, Black and McCool, 2001; Wilson, Fesenmaier, Fesenmaier and van Es, 2001).

Also from the demand side, farm tourism is characterized by several strengths involving the "intensity of the tourist experience" offered by this sector (Ingram, 2002). In contrast to mass tourism, farm holiday–makers often choose this type of tourism as a means of escaping from the city (Nickerson et al., 2001; Oppermann, 1996). In addition, there is also a strong motivation to see children learning from farm activities (Ingram, 2002) and increasing their knowledge about agriculture and the food chain (Nilsson, 2002; Sidali et al., 2007). Finally, farm tourism gains from the special appeal of rural areas because of the "mystique" associated with them (Wilson et al., 2001).

These reasons can explain the huge number of repeat guests that typifies farm tourism both in Germany and abroad (Embacher, 1994; Oppermann, 1995) and which has lead Busby and Rendle (2000) to declare that "the relationship between the host and guest is the main strength of farm tourism" (ibid, p. 640).

Weaknesses

Weaknesses have been identified as internal factors that can influence the success or competitive advantage of a business. They are under the control of the operators who need to improve them in order to surpass or at least reach other competitors' levels.

The lacking of a law entirely dedicated to farm tourism is a weakness of the sector which implies further problems in acquiring reliable statistics. Other weaknesses of the German farm tourism sector occur on the demand side. In contrast with other countries, such as Italy and France, where this tourist sector is linked with food traditions but not necessarily with low prices, in Germany the image of farm tourism tends to be that of a low-budget holiday (Nilsson, 2002). Further image deficits related to German farm tourism have been identified by scholars, including few activities for guests (Lehle, 1982; Pevetz, 1978), lack of comfort and some "hygiene" deficits (Pevetz, 1978, p. 13). Furthermore, Wagner et al. (1997) state that the reason the demand for farm tourism fell in the 1960s was its perception as a "primitive holiday" which was probably conveyed by the extreme age of the buildings and the generally low level of investment.

Also the traditional concept of farm tourism as a holiday (exclusively) for families with children has been considered a risky image for years. In fact, there is the risk of a target dependency (BLFUW, 2007; Lüdke, 2001), which is even more serious if we take into account the decreasing fertility rate in Germany (EUROSTAT, 2008).

Starting in the 1970s, the DLG (the major farm tourism association) has tried to replace the low-budget image of farm tourism with that of a "quality holiday for a fair price" by means of the DLG certification system (Wagner et al., 1997). As a consequence, in the last few decades, the image of German farm tourism has considerably improved (ibid.). However, in the collective imagery, farm tourism in Germany still remains a "families-with-children holiday". Furthermore, although the perception of farm tourism cannot be considered negative, it still lacks significant attractive power. Other sectors related to rural tourism, such as camping holidays or the international farm tourism destinations of Austria and Italy, attract German customers more successfully (BMELV 2006; BMELV, 2008). A study by Sidali and

Spiller (2008) conducted in Germany in 2007 on a sample of 567 respondents confirms the relatively poor attractive power of German farm tourism.

Another weakness related to German farm tourism and common to most tourist destinations is the high seasonality of the sector. According to the German government (BMELV, 2006, p. 16), more than the half of all farm holidays are taking between June and August.

A particular problem affecting this sector is the visibility of farm operators. According to Lemke (2003), the marketing strategies adopted by the associations of farm tourism operators could be improved if they were unified under the umbrella of a common institution operating at a national level, as it is the case with Austria (see Embacher, 1994, for a review of the Austrian farm tourism market). In Germany, farm operators are organized into several agricultural and tourist-based associations scattered over regional and national levels, and this dispersion works to the detriment of a coherent corporate identity (Lemke, 2003). Similarly, the co-existence of several certification systems remains an issue. Among them, the conventional star system provided by the national tourism board and the DLG certification of the German association for agriculture are the most widespread. Nevertheless, the message conveyed to the customer is confusing.

Another problem of farm tourism concerns investments. The findings of a study conducted by Sidali et al. (2007) show that investments in promotion are of paramount importance for the success of German farm tourism; however, only a small number of farm operators seem to be aware of this fact.

Thus, it appears that farmers are still lacking sufficient marketing knowledge in this sector because they rely heavily on their intuition. These findings are also confirmed by Busby and Rendle (2000, p. 638), who state that "many farmers are isolated with a lack of knowledge, expertise, and training in the tourism field".

2 State of the art of Italian farm tourism

2.1 Definition and typology

In Italy agritourism has "its own legal, administrative and fiscal structure" (Dettori, Paba and Pulina, 2004, p. 22). In fact, in this country agritourism is regulated by the law No. 96 of 20 February 2006, which replaced the outline law No. 730 of December 5, 1985.

Hence, the main feature of Italian agritourism is a rich legislative body dedicated to this sector which, since the majority of countries do not distinguish between rural tourism and agritourism, is unique in Europe (Di Muzio et al., 2000; INNOREF, 2006). In addition, the first Italian law on agritourism (730/1985) can be considered a pioneer of the European Agenda 2000 (ibid, p. 8) since it already contains several principles afterwards included by the European Union (for instance, measures for the sustainable development of rural areas and improvement of farmers' standard of living). In the following, we will focus on the development of the Italian regulation of agritourism.

Law No. 730/1985

The ultima ratio of this law is the principle of connection (*principio di connessione*) between farming activities and tourist activities. The latter encompass accommodation, catering, school-orientated farms, outdoor activities and the like, and are legally only allowed to be a secondary source of income. Thus, Italian agritourism can only take place on working farms, and the host has to be a farmer (pursuant to Article 2135 of the Civil Code) or a form of joint-stock companies or partnerships (INNOREF, 2006, p. 28). Moreover, in order to use the name *agriturismo* (i.e., agritourism), the agricultural activities of the farm, and not its tourism activities, must be predominant. However, this proportion is not fixed in terms of income but in terms of working hours. According to Carbone and Ribaudo (2000), this measure is very effective, boosting agritourism activity in less developed regions and supporting agriculture in richer agricultural areas.

In his study of 1994, Gregori draws the state of the art of Italian farm tourism and criticizes some problems related to the law No. 730/1985. For instance, he stresses that the principle of connection as stated in general terms by the law has led Italian regions to translate it in general terms as well. For instance, they have just introduced a threshold in the bed capacity of farms as a criterion for receiving the financial subsidies available to start-up agritourism operations. Several other restrictions, such as the predominance of farm-made foods over externally sourced food products, can also be traced back to the above-mentioned law. Furthermore, according to Gregori (1994), this law makes it impossible for farmers to grow and reach the size of other tourist operators. Some years later, other scholars report that Italy has a very limited number of agritourism operations in comparison with other European countries and that this is a consequence of the strictness of this law (Di Muzio et al., 2000; Idda et al., 2001).

Law No. 96/2006

Although in many aspects the law No. 730/1985 remains "an organic reference on the agritourism activity" (Agostini, 2007, p. 4; Russo, 2007, p. 75), the Italian legislator in 2006 is forced to promulgate law No. 96. This law provides that the tourism activities of the farm be connected to agriculture, which remains the fundamental role of the farm. However, the value of this new law is that it boosts the pluriactivity of farms by enhancing the role of all activities related to farming (such as the provision of hospitality) and thus contributes to their success. In gastronomy, for instance, the rule requiring the prevalence of farm-produced foods over external products has been replaced by a rule mandating the prevalence of locally produced food products. Furthermore, in order to promote Italian regional gastronomic traditions the law recommends the provision of foods with protected quality names such as PDO and PGI[5] (ARM, 2007), which are so-called niche commodities, for they are produced on a small-scale and in very limited areas (Belletti et al., 2007, p. 520).

---

[5] For an overview of protected quality names for agricultural products and foodstuffs see reg. (EEC) no. 509 and no. 510 of June 20, 2006, as well as the database of protected food product names in the European Union: http://ec.europa.eu/agriculture/quality/door/browse.html?display

Mention should also be made of those activities that are not directly bound to agriculture but are nevertheless essential for adding value to farm tourism, like wellness services. In this case, the new law establishes that they must be for the use of the farm's guests.

Thus, it is possible to define Italian agritourism in a number of ways. The first is to refer the relevant laws, as discussed above. Next, depending on the scope of their studies, some authors use geographical location to distinguish among farm operations. Then, Marino, Mastronardi and Rubertucci (1999) use the principle of connection as their point of departure. Table 5 offers an overview of the different approaches.

Table 5. Criteria for the definition of farm tourism

Defining criteria	Example
Legal framing	National level: Law No. 730, 1985 Law No. 96, 2006 Regional laws
Location	Mountain Hills Sea Cultural cities
Principle of connection	1. Less agriculture and high agritourism development 2. High agriculture and high agritourism development 3. High agriculture and less agritourism development 4. Low agricultural and high agritourism development

Source: own representation based on Di Muzio et al., 2000; Lo Surdo, 1988 and Marino et al., 1999

2.2 Historical development

Lo Surdo (1988) identifies three historical steps that explain the creation of agritourism.

In the first half of the twentieth century, urbanization and industrialization processes take place simultaneously. Nevertheless, the main values of agrarian society remain untouched.

Between the 1950s and the oil crisis of 1973–1974, increasing industrialization is accompanied by negative repercussions on society and the environment, such as uncontrolled urbanization and a social identity crisis, resulting in a fracture between agrarian and urban societies. The subsequent arrest of the industrialization process due to the oil crisis causes a generalized sense of insecurity. Individuals start looking back to the traditions associated with the agrarian society of the past. Already in 1975, Magagnotti suggests that agritourism could function as a means of "re-establishing the equilibrium between nature and the industrialized society" (ibid, p. 13). In these years, the first farmers of the region of Trentino Alto-Adige begin to transform their farms into agritourism accommodation influenced by the Austrian and German models (Garruti et al., 2003, p. 314).

The 1980s are characterized by a real "come back to the countryside" sentiment. Many city dwellers return to the countryside, at least during their holidays; others buy and refurbish old houses and use them as second homes. There is an increase in enrolment at agricultural colleges and universities, and a growing number of city dwellers show a preference for buying products at farms rather than supermarkets. During this period, the first agritourism operations in central and northern Italy open their doors to tourists and experience rapid growth.

According to a number of authors (Flabiano and Di Santolo, 2001a; Paolini, 2000), the first farmers to decide to go into this business are winemakers. However, in some regions, social reasons also boost the creation of this tourism form. For instance, in Sardinia in 1977, the first agritourism operations are founded in order to empower women in rural areas (Idda et al., 2001, p. 37). Similar findings are described by Flabiano and Di Santolo (2001b), who focus on agritourism in the Italian region of Friuli Venezia-Giulia. The institutionalization of this sector follows a similar path. According to Germini (1990) this process is boosted by farmers and men of culture who had examined and studied the phenomenon of agritourism in different European countries. In 1968 the first conference on agritourism, *Città e campagna* (City and countryside), takes place in Florence (Lo Surdo, 1988, p. 30) with the stated objective of protecting gastronomic traditions and the agrarian landscape through agritourism. As a consequence, the first association dedicated to the development of agritourism, Agriturist, is founded by the Confagricoltura (Italian Farmers' Association). In the following years, the other two main agritourism associations are also founded: Terranostra (controlled by the Italian Farmers' Association Coldiretti) in 1972 and Turismo Verde (the former Alturist) (controlled by the CIA, Confederazione Italiana Agricoltori) in 1975.

According to Di Muzio et al. (2000), the 1980s can be considered the official "birth decade" of this tourist sector, which is rather late in comparison to the northern European countries (Naspetti, Segale and Zanoli, 1999). Since then, Italian farm tourism has undergone rapid growth. For instance, between 1997 and 2004, it has displayed a growth of 65%, from 8,034 agriturismo facilities in 1997 to 14,017, in 2004 (Adua, 2007).

2.3 The supply side

In Italy farm tourism presently includes around 17,895 farm tourism operators (Agriturist, 2008). This figure refers to the operations allowed to incorporate the label *agriturismo* in their trade names. Among these, 14,810 are authorized to provide lodging for holidaymakers.

Table 6. Supply side of Italian farm tourism

Supply side	
Market share	0.7%[6]
Total number of farm tourism operations	17,895
Total number of farm operations with lodging permission	14,810
Total revenue (millions €) (2008)	1,008
Share tourist revenue : total revenue	30-40%
Average No. beds per farm	18
Participation of female farm operators	High
Mean age of farm operators (years)	48
Occupancy rate[7] (nights per year)	Above 200
Location	
Mountain	50%
Hills	37%
Plain	13%
Other competitors	
Market share of hotels in Italy	16%[8]
Market share of the camping sector in Italy	1.9%[9]
Market share agritourism in relation to hotels	4.3%
Market share agritourism in relation to camping	37.3%

Source: Adua, 2007; Agriturist, 2008 www.agriturist.it [November 27, 2008]; ARM, 2003; Flabiano and Di Santolo, 2001b and ISTAT, 2004

[6] Calculated as follows: ratio of the total revenue of Italian agritourism € 1,008,000,000 according to Agriturist (2008) to the total revenue of the Italian tourist sector of € 140,000,000,000 according to Federalberghi (2008) ([http://www.federalberghi.it/notizia1.asp?id=4723] November 2008)

[7] Total number of nights per year during which beds are occupied

[8] Calculated as follows: ratio of the total revenue of the Italian hotel sector € 23,000,000,000 according to by Databank (Databank Consulting) ([http://www.databank.it/index/index2.html] November 2008) to the total revenue of the Italian tourist sector € 140,000,000,000

[9] Calculated as follows: ratio of the total revenue of the Italian camping sector € 2,700,000,000 according to Databank (Databank Consulting) ([http://www.databank.it/index/index2.html] November 2008) to the total revenue of the Italian tourist sector € 140,000,000,000

On the other hand, farm operators with lodging permits can also have a second specialization, such as catering or outdoor activities. According to Agriturist (ibid.), the estimated yearly turnover of Italian agritourism in 2008 was around € 1,008 million, which corresponds to an increase of 4.4% compared to the previous year. Around one-third of all farm tourism operations are managed by women (Adua, 2007).

However, since the number of farm operations has increased as well, the average turnover per operation has decreased by approximately - 2% (AgE, 2007b). The average percentage of total revenue made up by agritourism is between 30% and 40% (Idda et al., 2001, p. 68). The regions of Tuscany, Sicily and Lombardy score the highest number of guests overall (both national and international), whereas the Venetian region, Liguria and Sardinia host the highest concentration of international guests (ibid.). Tuscany continues to be the region with the highest number of available beds, while Alto Adige is the region with the highest number of operations (Adua, 2007; Becheri and Bartolini, 2000; Lo Surdo, 1988). In their study on Sardinian agritourism, Idda et al. (2001, p. 67) state that there is a strong relationship between this tourist form and farming patterns. The majority of farm operations have a varied agricultural system, and crops are more common than animal husbandry. In the case of the latter, small animals are preferred, primarily for catering purposes. As mentioned above, Agriturist, Terranostra and Turismo Verde are the main associations responsible for promoting this tourist sector (see Figure 1)[10]. Since 1981 the three agritourism associations have joined the group Anagritur, which was founded in order to improve political and lobbying activity on behalf of agritourism. It also promotes further studies on agritourism-related issues.

Figure 1. Italian agritourism associations

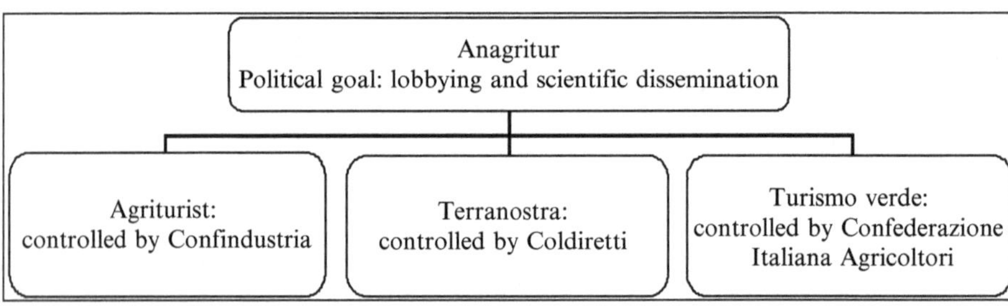

Source: own representation

Agriturist, Terranostra and Turismo Verde are all backed by the national agricultural associations, and they support farmers by listing their addresses in brochures, by allowing them to use booking facilities on their homepages and by creating alliances with partners located abroad. Despite this, some farmers do not judge the work of the main agritourism association very positively (Marinelli, 2001), which may explain

[10] Most farmers of South Tyrol belong to "Roter Hahn" which is controlled by Bauernverband Südtirol

why many of them prefer to join local pools of operators. The degree of fragmentation in the sector is therefore quite high.

2.4 Market demand and potential

In the following section, the most important traits of Italian farm guests are presented based on both governmental and academic literature. Among the most recent studies, the survey carried out by Coldiretti in 2008 (www.coldiretti.it [December 5, 2008]) is a good point of departure. The findings stress that, out of 1,700 individuals, around 3% are repeat farm guests, 4.2% occasional farm guests and 5.5% potential farm guests. Around 90% of respondents are not interested in farm holidays. The majority of farm guests are aged between 35 and 54 (41%) and have an intermediate to high education level, 36% of repeat guests are aged between 18 and 34, and 23% of repeat guests are over 55. The high education level of farm guests is confirmed by previous studies (Gregori, 1994; Naspetti et al., 1999). With a monthly income of around €2,266, the Italian farm holidaymaker displays a higher income than the national average of € 1,838.

As also confirmed by Cambi in his study of 2006 (http://www.centoare.it/ricerche_studi.asp [November 27, 2008]) and by Naspetti et al. (1999), it seems that the main target of Italian agritourism are families with parents aged between 30 and 40, although younger people and single people are also represented. With regard to the latter, they usually prefer to spend a farm holiday in a group. The low propensity to spend a farm holiday among individuals traveling alone has also been noted by previous studies (Naspetti et al., 1999). Many scholars (ARM, 2003) refer to new consumer trends related to Italian farm tourism, such as the recent tendency of young people to spend New Year's Eve on farms (ibid.). Finally, the reduced number of business trips to farms seems to confirm the fact that farm tourism is still not achieving its real market potential in this segment (see Table 7).

Table 7. Composition of Italian demand of agritourism

	Families	Couples	Groups	Singles	Business[a]	Business[b]
Agritourism	47.8%	43.1%	2.9%	3.5%	2.3%	0.3%
Italian tourist sector	37.7%	38.7%	6.4%	8.6%	8.5%	0.3%

[a] individuals [b] groups
Source: own elaborations

Table 8 gives more details on the composition of the demand for Italian farm tourism. Although the majority of farm guests are Italian, the percentage of foreign guests (primarily Germans) is also very important (69.8% and 30.2% respectively) (Cambi, 2006). In fact, according to the Italian Observatory on Tourism (ONT, 2008), foreign guests spend on average one day more on the farm. With a percentage of 38.5%, the number of guests who book a farm holiday on the Internet is quite high. Finally, Italian farm tourism also experiences a high number of repeat guests.

Table 8. Features of Italian agritourism demand

	Agritourism	Italian tourist sector
Italian guests	69.8%	71.3%
Foreign guests	30.2%	28.7%
Nights stay Italian guests	4.8*	4.4*
Nights stay foreign guests	5.9*	5.5*
Internet guests	38.5%	34.7%
Repeat guests	32.1%	42.0%

Source: Cambi, 2006; *ONT, 2008 (http://www.governo.it/Governo Informa/Dossier/rapporto_ turismo/index.html [January 16, 2009])

Overall, the findings of many scholars tend to confirm that the main motivations for spending a farm holiday are contact with nature, interest in agricultural life and traditions, need to escape day-to-day life and special interest in typical regional and/or organic products (Cambi, 2006; Idda et al., 2001; Marino et al., 1999). Other reasons, such as price convenience and curiosity, are not significantly represented (Gregori, 1994; Naspetti et al., 1999). Concerning guests' favorite activities, it seems that national and international guests show two tendencies: the majority of Italian farm guests use the farm as a point of departure for visiting the natural and archaeological surroundings and sampling food and wine specialties. On the other hand, international guests spend more time on the farm and make greater use of its leisure facilities (Lo Surdo, 1988).

Table 9. Demand side of Italian agritourism

Demand side	
Market share (farm tourism stays in relation to total overnight stays)	3.27%
Total number of guests in 2008 (millions)	2.1
Main traits of guests of Italian leisure farms	
Age (years)	41
Monthly income (in euros)	2,266
Education	Intermediate-to-high
Motivations for choosing a farm holiday	**Activities during a farm holiday**
1. Contact with nature	1. Visiting natural surroundings
2. Agricultural life and traditions	2. Visiting archaeological highlights
3. Escape day-to-day life	3. Tasting oeno-gastronomical specialties

Source: own representation based on AgE, 2008; Cambi, 2006; Coldiretti, 2008

2.5 Strengths and weaknesses of Italian farm tourism

Strengths

As with other forms of tourism located in the *Belpaese*, Italian farm tourism also benefits from the favorable location, dramatic surroundings and propitious climate. As reported by Cambi (2006), despite the adverse economic conjuncture "Italy still remains the country with the greatest evocative power over foreign vacationers". Furthermore, this tourist form is not perceived as the prerogative of a particular market segment as it attracts both families with or without children as well as young people traveling in groups (Cambi, 2006; Idda et al., 2001; Lo Surdo, 1988). A literature review identifies the following features as its main strengths.

A great architectonical variety of farmhouses: since farm buildings had to be adapted to regional climatic conditions, each geographical area in Italy has its own farmhouse style. For instance, in the Alps *masi altoatesini* are ancient farmsteads for cattle husbandry, in the Padan regions *case coloniche* and *casali* are farmhouses with a quadrangular form with or without a dovecot tower, in the south of Italy *masserie* are large estate properties with sumptuous decorations that used to belong to rich landlords, and *trulli* in the Apulia region are white houses with conical brown roofs.

A rich oenological and gastronomical heritage: if many Italian farm operators can boast of architectonical richness, their oenological and gastronomical traditions are even richer. According to Paolini (2000) many farm and rural holidaymakers are showing an increasing interest in rediscovering wine and food traditions. In this field, Italy has been a forerunner. With 109 and 165 food products respectively, Italy has the highest number of both PDO (Protected Designation of Origin) and PGI (Protected Geographical Indication) products of any European country (as of December 2007). Since their success is deeply-rooted in agrarian traditions, many farm guests associate them with the farm landscape, which is a great asset for the whole farm and rural sector.

Closely connected with the latter is, according to Ammirato (2007), "the ability of typical agrifood producers to access to Short agri-Food Supply Chains (SFSC) of which agritourism represents a significant example". In fact, "while it reduces the agrifood supply chain stages, [farm tourism] is able to be the contact point between the request for quality of life and the offer for typical products".

The image of Italian farm tourism has benefited a lot from the particular fascination of Tuscany, which has been carefully fostered by the local administration for many years (Balestrieri, 1997). In addition, this tourism form is attractive for both domestic and international tourists (Cambi, 2006; Idda et al., 2001, Lo Surdo, 1988).

Finally, as mentioned at the beginning of this section, agritourism has a particular legal status, which gives some fiscal advantages to this niche sector in comparison to other tourist sectors.

Weaknesses

One of the main weaknesses of Italian farm tourism lies with regional legislation dealing with the sector. Since Italy is a regionally administered country, there is no comprehensive legislative corpus for agritourism. Each region decides, based on the agricultural characteristics of the area, to what extent the general criteria stated in national laws should apply to local needs. For instance, Tuscany does not allow farmers to provide meals to any but their own residents, whereas in other regions, such as Umbria, it is permissible to provide meals to non farm guests. In their study on accuracy in interpreting national law to promote regional farm tourism, Idda et al. (2001; 2000) show that the northern regions of Friuli Venezia-Giulia and Veneto as well as the southern region of Abruzzo rank highest in providing detailed regional legislation dealing with farm tourism.

Another weakness is related to catering. Despite the legal recommendation in law 96 to use regional or certified foods in catering in order to promote the regional identity of the whole rural area (Brunori, 2003), Italian farm operators seem to have some difficulty adopting this "corporate view" although regional differences are huge. Many farmers underestimate the potential of promoting the area through gastronomic specialties and instead continue to provide standardized food to their guests (Idda et al., 2001; Lo Surdo, 2008). The same applies to regional handicraft products, which are often totally neglected by farm operators as well. This leads to further problems, such disappointment among farm guests who had anticipated a more regional cultural experience from their holidays. In general, Italian farm operators show an insufficient level of professionalism (ibid.).

The challenges that Italian farm operators have to cope nowadays differ from those in the past. Whereas in the 1980s the main concerns were quality related, such as the use of low-quality furniture or poor hygiene (Lo Surdo, 1988), the current supply of farm tourism does not differ very much from other rural tourism competitors. To these problems, Carbone and Ribaudo (2000, p. 69) add the illegal use of the label *agriturismo* by external tourism operators who are not legally authorized to use it.

Problems of minor magnitude facing farm operators include some image deficits. For example, many guests report becoming bored during their stay on the farm (Idda et al., 2001). This shows that farm operators have not invested enough in leisure activities, which seems invaluable, given the slow pace of rural life or, in Gregori's words, the "melancholy evoked by rural areas" (1994).

Another problem lies in Tuscany's powerful image. Admittedly, Tuscany's popularity has had positive repercussions throughout the sector, especially where prices are concerned. However, because this image does not suit their local identity, an increasing number of operators in other regions feel imprisoned by it (Lo Surdo, 2008).

Finally, insurance-related matters are also a problem. Unlike other tourist sectors, Italian farm guests come primarily to visit - and experience - the farm itself (cf. Ohe and Ciani, 2000). This means that many of them would like to take part in farming activities. However, the legal framework for doing so has not been sufficiently developed.

Table 10. Main strengths and weaknesses of German and Italian farm tourism

German farm tourism	
Strengths	**Weaknesses**
Political (national and European) support	Lack of legislation dedicated to farm tourism and of reliable statistics
Self commitment of farm operators	Lack of a common institution for unified marketing strategies (clear corporate identity)
High female participation	Co-existence of several quality certification systems
Environmental protection, biodiversity, etc.	Lack of professional skills of farm operators, especially in terms of marketing knowledge
Sustainable development of rural areas	Dependency on large-sized families
Social benefits (e.g., keeping farmers on the land, gender related benefits)	Lower attraction in comparison with other holiday types (city break, holidays abroad etc)
High customer loyalty	High seasonality
Italian farm tourism	
Strengths	**Weaknesses**
Architectonic agrarian heritage	Co-existence of different laws at the regional level
Linkage with food tourism (economies of scope)	Insufficient level of networking with other agents in the area
Smart image (Tuscany and varied target)	Low quality standards of activities programs
Short agri-Food Supply Chains	Tuscanization of farm tourism image
Fiscal advantages	Guests' expectations are easier to be disappointed than as in other hospitality sectors

Source: own elaborations

3 Conclusions and structure of the book

This brief analysis of farm tourism in Germany and Italy has introduced some issues related to this tourist sector that will be further developed in this book. In the first part, the German and Italian experience with farm tourism will be analysed using an interdisciplinary approach. Furthermore, the description of farm tourism in a Eastern European country (Romania) will be used to expand our knowledge on the complex nature of this tourist sector.

The second part of the book will elucidate the role of partnerships/networks with special interest actors - such as the Slow Food movement or educational farms. These contributions describe a range of products related to farm tourism (e.g., wine tourism, education) which constitute an enhanced supply of tourism in rural areas.

As stated by Roberts and Hall (2001) “evidence emerges of the ways in which new forms of consumption may alter the nature of supply [of farm/rural tourism]”. Accordingly, the central argument of the last part of the book is to appraise the potential of new avenues of research such as online and sensory marketing for fulfilling the needs of the new trends within this sector.

References

Adua, M. (2007). L'agriturismo nel suo piccolo cresce alla grande. *L'informatore agrario 12*(2007), 26-29.

AgE (2007a). Camping auf dem Bauernhof liegt voll im Trend. *Agra-Europe. Unabhängiger europäischer Presse- u. Informationsdienst für Agrarpolitik u. Agrarwirtschaft, 16*(07).

AgE (2007b). Agrartourismus in Italien weist Schwächen auf. Agra-Europe. *Unabhängiger europäischer Presse- u. Informationsdienst für Agrarpolitik u. Agrarwirtschaft, 51*(07), 38.

AgE (2008). Agrartourismus in Italien registriert Einbußen. *Agra-Europe. Unabhängiger europäischer Presse- u. Informationsdienst für Agrarpolitik u. Agrarwirtschaft, 37*(08), 4.

Agostini, S. (2007). Learning sustainability of rural tourism: farm competitiveness and landscape health risk assessment. *Agricultural Engineering International: the CIGR Ejournal, IX*(October), 1-11.

Agriturist (2008). www.agriturist.it accessed November 27, 2008.

Ammirato, S. (2007). The agritourism and the role of eCommerce in the agribusiness evolution: evidence from a regional survey. Proceeding of the EFITA Conference, July 2007, Glasgow (UK).

ARM (Azienda Romana Mercati) (Eds). (2003). Agriturismo and marketing. Roma: Agra.

ARM (Azienda Romana Mercati) (Eds). (2007). Voglio aprire un agriturismo. Roma: Agra.

BAG (Bundesarbeitsgemeinschaft für Urlaub auf dem Bauernhof / German association of farm tourism) (Ed.) (2008). Geschäftsbericht 2007/2008. BAG: www.bauernhofurlaub-deutschland.de/ accessed November 27, 2008.

Balestrieri, G. (1997) (Ed.), L'agriturismo in Toscana. Le imprese, gli ospiti, gli intermediari. Milano: Franco Angeli.

Barth, K., and Theis, H. - J. (1998). Hotel-Marketing. Strategien, Marketing-Mix, Planung, Kontrolle (2nd Edition). Wiesbaden: Gabler.

Becheri, E and Bartolini C. (2000). Le componenti del mercato nazionale e regionale: l'offerta turistica, in AA.VV. (Ed.), *Nono rapporto sul turismo italiano*, Rontassieve: Centro Stampa 2P.

Belletti, G., Burgassi, T., Manco, E., Marescotti, A., Pacciani, A., and Scaramuzzi, S. (2007). The role of geographical indications (PDO and PGI) on the internationalisation process of agro-food products. Paper presented at the International Marketing and International Trade of Quality Food and Products, Bologna.

BLFUW (Österreichisches Bundesministerium für Land und Forstwirtschaft, Umwelt und Wasserwirtschaft / Austrian Federal Ministry for Agriculture, Forestry, Environment and Water Management), (2007). Urlaub am Bauernhof aus der Sicht von Gästen und Anbietern. Retrieved, from: www.dafne.at, June 6, 2008.

BMELV (Bundesministerium für Ernährung, Landwirtschaft und Verbraucherschutz / German Ministry for food, agriculture and consumers' protection) (Ed.), (2006). Urlaub auf dem Bauernhof / Urlaub auf dem Lande. Studie für das Bundesministerium für Ernährung, Landwirtschaft und Verbraucherschutz auf der Basis der Reiseanalyse 2006 (Study for the BMELV on the basis of travel analysis 2006). Bonn.

BMELV (Bundesministerium für Ernährung, Landwirtschaft und Verbraucherschutz / German Ministry for food, agriculture and consumers' protection) (Ed.), (2008). Expertise "Urlaub auf dem Bauernhof / Lande". Bonn.

Bodenstein, G., and Spiller, A. (1998). Marketing. Strategien, Instrumente und Organisation. Landsberg/Lech: Moderne Industrie.

Brunori, G. (2003). Percorsi di convergenza tra nord e sud del mondo - Cibo, modelli di consumo e modelli di sviluppo. *Il Ponte: rivista di politica, economia e cultura, 59*(6), 139-152.

Busby, G., and Rendle, S. (2000). The transition from tourism on farms to farm tourism. *Tourism management, 21*(6), 635-642.

BWT (Bundesministerium für Wirtschaft und Technologie / National Ministry of Economics and Technology) (Pub.) Tourismuspolitischer Bericht der Bundesregierung. 16. Legislaturperiode. (2008). (T. Deutschland / Bundesministerium für Wirtschaft und Technologie). Berlin.

Caballé, A. (1999). Farm tourism in Spain: a gender perspective. *Geojournal, 48*(3), 245-252.

Cambi, C. (2006). Gli Italiani e le vacanze in Agriturismo. Paper presented at the Agri&tour, Arezzo (Italy) November 17-18, 2006, from www.centoare.it/ricerche_studi.asp retrieved November 27, 2008.

Carbone, F., and Ribaudo, F. (2000). Agriturismi e politiche di sviluppo. *Rivista di politica agraria, 3*(3), 47-73.

Coldiretti (2008). www.coldiretti.it accessed November 27, 2008.

Dettori, D. G., Paba, A., and Pulina, M. (2004). European rural tourism: agrotouristic firms in Sardinia and their life cycle. Unpublished manuscript, CRENOS.

Di Muzio, R., Dota, S., Faggioli, F., Manilla, M., and Tibiletti, E. (2000). Agriturismo e sviluppo delle aree rurali. Bologna: Calderini Edagricole.

Embacher, H. (1994). Marketing for agri-tourism in Austria: strategy and realization in a highly developed tourist destination. *Journal of Sustainable Tourism (UK), 2*(1/2), 61-76.

EUROSTAT (2008). Demographic outlook. http:// epp.eurostat.ec.europa.eu/cache/ITY_OFF PUB /KS-SF-08-081 /EN/KS-SF-08-081-EN.PDF accessed December, 10 2008.

Flabiano, C., and Di Santolo, S. (2001a). Agriturismo in Friuli-V.G. analisi dell'offerta e della domanda. 1. Offerta agrituristica: evoluzione e caratteristiche della componente umana: Ersa.

Flabiano, C., and Di Santolo, S. (2001b). Agriturismo in Friuli-V.G. analisi dell'offerta e della domanda. 2. Offerta agrituristica: l'aspetto agricolo e le trasformazioni: Ersa.

Garruti, C., Clavai, P., Pagnini, M. P., and Scaini, M. (2003). The farm-holidays: a diversification attempt of the tourist supply. Calabria event. Paper presented at the Conference The Cultural Turn in Geography, 18-20th of September 2003 - Part VI: Tourism, Sustainable Development and Culture Turn Gorizia.

Germini, C. (1990). Agriturismo a contatto con natura e ambiente, Bologna, Calderoni.

Gregori, M. (1994). Agriturismo, turismo rurale e sviluppo rurale. Paper presented at the XXXI convegno di studi della Sidea, Campobasso.

Idda, L., Benedetto, G., Pulina, P., and Rubino, C. (2001). Aspetti territoriali dell'offerta agrituristica in Sardegna. In L. Idda (Ed.), *Modelli di offerta e politiche di governo dell'agriturismo in Sardegna* (pp. 13-86). Sassari: Gallizzi.

Idda, L., Furesi, R., Albino Madau, F., and Sini, M. P. (2000). Le politiche per lo sviluppo dell'agriturismo: funzioni ed efficacia della legislazione regionale. *Rivista di politica agraria, 55*(3), 329-366.

Ingram, G. (2002). Motivations of farm tourism hosts and guests in the South West Tapestry Region, Western Australia: A phenomenological study. *Indo-Pacific Journal of Phenomenology 2*(1), 1-12.

INNOREF. (2006). Agritourism, typical, organic and traditional products in Umbria, unpublished work.

ISTAT (2004) Dati annuali sull'agriturismo (Dataset December 31, 2003). www.istat.it [November 27, 2008].

ISTAT (2008) Dati annuali sull'agriturismo (Dataset December 31, 2007). www.istat.it [March 16, 2008].

Johnson, G., and Scholes, K. (1997). Exploring corporate strategy (4. Ed.). London [u.a.]: Prentice Hall.

Klöpper, R. (1974). Die räumliche Struktur des Angebots von "Urlaub auf dem Bauernhof". Entwicklungschancen im Rahmen des gesamten Beherbergungsanbebots in Landgemeinden (L. Akademie für Raumforschung und, Trans.). Bonn-Bad Godesberg: AID.

Lehle, K. (1982). Urlaub auf dem Bauernhof - Analyse und Perspektiven des Nebenbetriebes 'Tourismus in der Landwirtschaft'. Frankfurt M.: V. Kommentator.

Lemke, S. (2003). Marktanalyse Urlaub auf dem Bauernhof 2002. Ergebnisse der Direktbefragung (L. Bayerische Landesanstalt für and E. Bayerische Landesanstalt für, Trans.). Freising: Bayerische Landesanstalt für Landwirtschaft (LfL).

Lo Surdo, G. (1988). Agriturismo. Risorse-gestione-programmazione-esperienze. Bologna: Edagricole.

Lo Surdo, G. (2008). Management dell'azienda agrituristica. Public lecture of the Agri&tour Master, November 14, 2008. Arezzo (Italy).

Lüdke, M. (2001). Nutzung des Internet als Marketinginstrument für landwirtschaftliche Unternehmen. Ansatzpunkte für einen Online-Marketing-Mix landwirtschaftlicher Direktvermarkter sowie Anbieter von Urlaub auf dem Bauernhof. Aachen: Shaker.

Marinelli, H. (2001). Prodotti tipico-tradizionali e agriturismo nel Montefeltro: una valutazione sulle opportunità di sviluppo di un territorio. Alma Mater University, Bologna.

Marino, D., Mastronardi, L., and Rubertucci, M. G. (1999). L'agriturismo nella regione Molise: strategie aziendali e percorsi di sviluppo locale. *Genio rurale*, 11, 48-62.

McGehee, N. G., and Kim, K. (2004). Motivation for agri-tourism entrepreneurship. *Journal of travel research, 43*(2), 161-170.

Mook, R. (1978). Bauern als Unternehmer im Camping-Tourismus von Oberbayern. Farmers as entrepreneurs in the camping tourism of Upper Bavaria. *Norsk geografisk tidsskrift, 32*(4), 181-190.

Naspetti, S., Segale, A., and Zanoli, R. (1999). La domanda nazionale di agriturismo: risultati di un'indagine empirica. *Genio rurale*, 11.

Nickerson, N. P., Black, R. J., and McCool, S. F. (2001). Agritourism: motivations behind farm-ranch business diversification. *Journal of travel research, 40*(1), 19-26.

Nilsson, A. P. (2002). Staying on farms - an ideological background. *Annals of tourism research, 29*(1), 7-24.

Ohe, Y., and Ciani, A. (2000). On-farm tourism activity and attitudes of the operators: a Hiroshima-Umbria comparative case study. Tech. Bull. Fac. Hort. Chiba. University.

ONT (2008) www.governo.it/ GovernoInforma/ Dossier/ rapporto _ turismo / index.html [January 16, 2009]

Oppermann, M. (1995). Holidays on the farm: a case study of German hosts and guests. *Journal of travel research, 34*(1), 63-67.

Oppermann, M. (1996). Rural Tourism in Southern Germany. *Annals of tourism research, 23*(1), 86-102.

Paolini, D. (2000). I luoghi del gusto: cibo e territorio come risorsa di marketing. Milano: Baldini Castoldi Dalai.

Pevetz, W. (1978). Struktur und Motive von Urlaubern auf Österreichischen Bauernhöfen und deren Einstellung zur Landwirtschaft = Structure and motivations of holidaymakers on Austrian farms and their attitude towards agriculture". Schriftenreihe des Agrarwirtschaftlichen Institutes des Bundesministeriums für Land- und Forstwirtschaft, 27, Vienna.

Przezbórska, L. (2003). Relationships between rural tourism and agrarian restructuring in a transitional economy: The case of Poland. In: D. Hall, M. Mitchell, and L. Roberts (Eds.), *New Directions in Rural Tourism* (pp. 203-222). Aldershot: Ashgate.

Roberts, L., and Hall, D. (2001). Rural tourism and recreation. Principles to practice. Oxford [u.a.]: CABI Pub.

Russo, L. (2007). Agriturismo e impresa agricola. In Accademia dei Georgofili (Ed.) *L'agriturismo fra Stato, Regioni e Comunità europea*, (pp. 69-88). Firenze: Società Editrice Fiorentina.

Schulz-Borck, H., and Tiede, S. (1974). Urlaub auf dem Bauernhof. Analyse d. Angebots an Urlaubsquartieren auf Bauernhöfen (Kataloganalyse): Auswertung d. Antworten v. Urlaubern (Postkartenanalyse). Frankfurt a.M.: Verl. Kommentator.

Sidali, K. L., and Spiller, A. (2008). Factors influencing the image of German farm tourism: a path model approach. *Journal of Global Marketing Science, 18*(4), 35-61.

Sidali, K. L., Schulze, H., and Spiller, A. (2007). Farm vacation tourism between tradition and innovation. *Agrarwirtschaft und Agrarsoziologie, 2007*(1), 75-90.

Skuras, D., Petrou, A., and Clark, G. (2006). Demand for rural tourism: the effects of quality and information. *Agricultural economics, 35*(2), 183-192.

Tiede, S., and Schulz-Borck, H. (1975). Urlaub auf dem Bauernhof. Untersuchung des Angebotes und der Auslastung der Beherberbungsbetriebe (E. Bundesministerium für and V. Land- und Hauswirtschaftlichen Auswertungs- und Informationsdienst e, Trans.). Münster-Hiltrup: Landwirtschaftsverlag.

Wagner, P., Burger, H.-G., and Magnus, R. (1997). Ferien auf dem Bauernhof: Was wünschen die Urlauber? Ergebnisse einer Befragung. Frankfurt am Main: DLG-Publ.

Wilson, S., Fesenmaier, D. R., Fesenmaier, J., and van Es, J. C. (2001). Factors for success in rural tourism development. *Journal of Travel Research, 40*(2), 132-138.

Perspectives of emotional food communication for farm operators

Nina Stockebrand
Katia Laura Sidali
Achim Spiller

University of Göttingen
Department of Agricultural Economics and Rural Development
nstocke@uni-goettingen.de
{Katia-Laura.Sidali, a.spiller}@agr.uni-goettingen.de

Abstract

Similarly to other countries, farm tourism operators in Germany have also acknowledged the great potential of high-quality food (regional, organic, etc.) in reinforcing guests' ties to the territory and access, in this way, new target segments. However, due to limited entrepreneurship capabilities farmers may fail at finding the adequate communication of the food component within the farm tourism supply. In the marketing and tourism literature, emotional communication strategies, above all storytelling, are considered very successful. However, there is no empirical evidence whether a story or a more informative communication style is more successful in marketing of regional food in tourism. Hence, by means of an experimental design, this study compares different communication tools (information list, text and story) in order to detect within a sample of 122 students, which is the most appropriate strategy to promote high-quality food. Although storytelling shows the strongest emotional linkage with the readers, there is no evidence that this leads also to a higher buying intention as only 30 % of the respondents are convinced to buy the product after reading the story. In contrast, a high buying intention is reached by means of the text. Such a communication technique seems appropriate to promote regional food since it is not too emotional as to frighten individuals, nor too sparse as to reduce the attractiveness of the food.

Zusammenfassung

Ähnlich wie in anderen Ländern haben die deutschen Anbieter von Urlaub auf dem Bauernhof das große Potenzial von hochwertigen Lebensmitteln (regionale, bio, usw.) erkannt, um die Beziehung zwischen Gast und Land zu stärken und dadurch neue Zielsegmente anzusprechen. Jedoch können viele Landwirte aufgrund ihrer bisher begrenzten Marketingkenntnisse den Kunden das Lebensmittelangebot nicht hinreichend vermitteln. In der Marketing- und Tourismusliteratur werden emotionale Kommunikationsstrategien, vor allem das Geschichtenerzählen, als sehr erfolgreich angesehen. Es gibt aber keine empirischen Belege dafür, ob bei der Vermarktung regionaler Lebensmittel eine Geschichte oder ein eher informativer Kommunikationsstil erfolgreicher ist. Folglich wurden mittels eines experimentellen Designs verschiedene Kommunikationstexte verglichen (Informationsliste, Fließtext und Geschichte), um die beste Kommunikationsstrategie für hochwertige Lebensmittel zu ermitteln. Die Stichprobe in dieser Pilotstudie umfasst 122 Studenten. Obwohl

K.L. Sidali et al. (eds.), *Food, Agri-Culture and Tourism: Linking local gastronomy and rural tourism: interdisciplinary perspectives*, DOI 10.1007/978-3-642-11361-1_2,

das Geschichtenerzählen die stärkste emotionale Bindung bei den Lesern aufweist, gibt es keinen Nachweis, dass dies auch zu einer höheren Kaufbereitschaft führt. Insgesamt gaben aber 30% der Befragten an, das Produkt nach dem Lesen der Geschichte kaufen zu wollen. Im Gegensatz dazu ist durch den Fließtext eine hohe Kaufbereitschaft erzielt worden. Eine solche Kommunikationsmethode scheint geeignet, regionale Lebensmittel zu fördern, da sie weder zu emotional noch zu informationslastig ist.

Riassunto

Conformemente alle altre nazioni, anche in Germania gli operatori agrituristici hanno compreso il potenziale del cibo d'alta qualità (regionale, biologico, etc.) capace di rinforzare i legami degli ospiti con il territorio nonché di attirare nuovi turisti. Ciononostante, a causa di scarse abilità imprenditoriali, i gestori agrituristici non riescono sempre a comunicare efficacemente la componente agroalimentare che caratterizza l'offerta agrituristica. Nella letteratura turistica e di marketing sono considerate di grande successo le strategie di comunicazione emozionale quali, ad esempio, lo 'story-telling'. Tuttavia non c'è evidenza empirica sul fatto che una storia o una forma di comunicazione più informativa sia anche più efficace nella commercializzazione di prodotti alimentari regionali nell'ambito del turismo. Quindi sulla base di un disegno sperimentale, lo studio pilota esaminato in questo articolo confronta vari sistemi di comunicazione (un mero elenco di informazioni, un testo e una storia), per scoprire, su un campione di 122 studenti, quale sia la strategia più adatta a promuovere alimenti d'alta qualità. Anche se la tecnica dello 'story-telling' dimostra il più forte legame emotivo con i lettori, non c'è evidenza che ciò conduca pure ad una maggiore intenzione d'acquisto, dato che solo il 30 % di chi ha risposto intende comprare il prodotto dopo aver letto la storia. Invece l'intenzione d'acquisto è maggiore dopo aver letto le informazioni contenute nel testo. Quest'ultima sembra la tecnica di comunicazione più adatta a pubblicizzare il cibo regionale dal momento che non è fortemente emotiva.

Keywords: Storytelling; Farm tourism; Emotional food consumption.

1 Introduction

Farm tourism is a niche product in the tourist industry and is defined by Przezbórska (2003) as "all tourism and recreation activities connected with a working farm". Originating from the need of diversifying agricultural business (Sharpley and Vass, 2006), this tourist form has developed very differently among the European countries. For instance in Italy or France, farm tourism owes a great deal of its popularity to the close relation to the territory, which is transmitted to guests in manifold ways such as through the rediscovering of agrarian handicraft traditions, through the preparation of vernacular food, through the participation in village festivals, etc.

In contrast, in Germany the farm tourism sector is still not very well known among the population and only recently have farm tourism operators and associations tried to give more visibility to the sector in the hope to broaden their market segments (BAG, 2008; BMELV, 2006 and 2008). Similarly to other countries, farm operators in Germany have also acknowledged the great potential of regional food in reinforcing guests' ties to the territory, which has led to a proliferation of farm cafés, restaurants, sampling rooms, etc. (BAG, 2008). However, since many German farmers have limited entrepreneurship capabilities, the decision to include the food component within the farm tourism supply may challenge them in terms of, above all, adopting the most appropriate advertising strategy.

As demonstrated by previous research (Clarke, 1996), one of the main weaknesses concerning farm tourism facilities is communication. Surprisingly, the extant literature demonstrates limited interest in this topic, although it has been recognized that food poses problems when it is positioned in the general context of tourism (Cohen and Avieli, 2004; Hashimoto and Telfer, 2006; Telfer and Wall, 1996) or rural tourism (Murray, 2005).

Since tourism is an experience product, it is plausible to suppose that emotional communication techniques are a strong driver for the promotion of those farm tourism operations which are specialized in (or eager to specialize in) high-quality catering. Hence, the main emphasis of this study is set on storytelling, which has been identified by Dehuang, Lixin and Woodside (2009) as the emotional communication technique that provides "creative clues for positioning a destination uniquely and meaningfully in the minds of potential future visitors."

On the other hand, the prosaic nature of food implies that individuals cope with the decision to ingest something which is very intimate and personal (Cohen and Avieli, 2004). This may require a communication technique which vehicles familiar and "tangible" features rather than emotional, creative and "intangible" ones. As a consequence, our research objective is to identify the most appropriate communication technique for farm operators in Germany to promote regional high-quality food. This is a pilot study with an explorative approach. By means of an experimental design, this paper compares different communication tools (text, story and information list) in order to promote regional food.

2 Farm tourism in Germany

The origin of farm tourism in Germany dates back 150 years (Nilsson, 2002), but it was after the Second World War that this sector began to gather momentum (Wagner, Burger, and Magnus, 1997). Farm tourism boomed in the 1970s (Oppermann, 1996); however, it was only during the 1990s that this tourist sector really gained national visibility. German farm tourism reached its peak in 1996 with 3.9 million guests. It dropped again in the period 1999-2001 before gradually recovering. According to the German Ministry for Food, Agriculture and Consumer Protection (BMELV, 2006), there are almost 25,000 farm-based tourist facilities in Germany, which generate an annual turnover of 943 million Euros.

Because of its distinctive traits such as child-friendliness, natural settings, leisure, etc. (Embacher, 1994; Oppermann, 1995 and 1996; Schöppner, 1988; Wohlmann and Lohmann, 1986), farm tourism has established itself as a vacation particularly suitable for families with children (BMELV, 2008). Moreover, because most of the supply is still limited to the provision of basic services (e.g., catering is reduced to breakfast, accommodation is simple) this tourist form is considered a low-budget holiday (Nilsson, 2002). Against this background, it is not surprising that farm tourism in Germany seldom enters in the choice set of young individuals or couples without children when they are planning a holiday.

The crystallization of these traits over the years has caused a high dependency of the sector on the family target, which is particularly risky since this is a shrinking market segment both in Germany and in Europe. This situation is compounded by the high level of competition in the rural tourism sector: as stressed in many studies, the attractive power of farm tourism in Germany has been decreasing in the favor of other sectors related to rural tourism, such as camping holidays (BMELV, 2008). Even more surprisingly, many Germans who would not plan a farm holiday in their country are attracted by farm tourism destinations when they travel abroad; e.g., the international farm tourism destinations of Austria and Italy (BMELV, 2006). This may be explained by the fact that, among other attractions, in these countries the farm holiday is successfully linked with local and traditional food (Embacher, 1994; Brunori et al., 2003). In Germany, the process of rediscovering food-related traditions has only recently started as confirmed by the low number of registered PDO (Protected Designation of Origin) and PGI (Protected Geographical Indication) food labels (Spiller, Voss, and Deimel, 2007). Despite this, there are notable examples of culinary heritage such as "Heidehönig" (heather honey from the Lüneberg Heath) and "Leinelamm" (lamb from the rare breed of Leineschaf sheep).

Among the generally rather undefined supply of German farm tourism, there are some exceptions of farm tourism operators who have distinguished their supply by adding value to their catering with high-quality oeno-gastronomical products such as organic food. There are about 18,703 organic farms in Germany (Oekolandbau, 2008), which are organized in organic farming associations (Bioland, Demeter, etc.), but only approximately 400 offer tourist services. However, this number is estimated to increase further due to the growing appreciation of organic products displayed by Germans (Eghbal, 2008). In fact, in the past years, the market of certified organic products has been constantly increasing as a reaction to the recurrent food scandals which have taken place in the country (Albersmeier, Schulze, and Spiller, 2007).

It is the intention both of the German farm tourism associations in general and of the German government to boost farm tourism in Germany and to strengthen the linkages with high-quality food production (e.g., local, organic, etc.; BAG, 2008).

3 Farm tourism, food and emotions

3.1 Product development strategy based on high quality food

Efforts to readjust the image of the farm tourist sector as a quality holiday for a fair price have already been carried out in the 1990s (cf. Wagner et al., 1997); however, it is especially in the last years that the main farm tourism associations have encouraged their members to differentiate their supply in order to access new tourist segments (cf. BAG, 2008). The extant literature suggests several product development strategies for farm tourism operations (Adams, 2008; Williams et al., 2001), ranging from the connection to rural festivals to the provision of leisure activities such as hunting, fishing, etc.

Especially the inclusion of catering based on high value foods (regional, geographically protected, organic, etc.) could be a sound strategy in order to access new tourists. First of all, farm tourism operations could attract gourmet or culinary tourists, who are defined as those tourists whose primary intention to visit a place is connected to taking part in a culinary experience (Murray, 2008). Not only gourmets but also the so-called LOHAS (Lifestyle of Health and Sustainability) consumers could show interest in those farms which position themselves as venues of high-quality food if this is perceived as authentic, customized and convenient (Kuhl and Niessen, 2009).

A look outside Germany reveals that some countries have been particularly successful in integrating the food component in the domestic supply of farm tourism. In Italy, for instance, the consumption of high-quality food has turned into one of the main attractions of farm tourism. This has been possible thanks to introduction of the Italian Law No. 96/2006, which disciplines the farm tourism sector and explicitly mandates the prevalence of locally produced food products. Hence, in order to promote Italian regional gastronomic traditions, the law recommends the provision of foods with protected quality names such as PDO and PGI (cf. ARM, 2007), organic or vernacular food.

In Germany, some farm tourism operators are diversifying their supply by offering culinary-oriented services such as farm cafés, farm-based restaurants, wine-tasting cellars, etc., where regional high-quality food is offered not only to over-night guests but also to the increasing number of one-day farm guests as well (BAG, 2009).

A growing number of farm tourism operators have been attempting to distinguish themselves as trustworthy partners in the food chain and they have signaled it to the new target segments by offering high-quality catering. This is of paramount importance since it is estimated that more than one-third of tourist expenditure is on food (Telfer and Wall, 1996). However, this implies the existence of two prerequisites. Firstly, an appropriate assortment of regional food specialties, the discussion of which would go beyond the scope of the present study. Secondly, a convincing communication strategy.

The latter is not easy to set up. Due to the limited marketing knowledge of farm operators and the lack of studies which deal with the food component in (farm) tourism we decided to devote the following to the analysis of successful communication strategies for farm tourism operators who want to promote regional food products to farm guests. The purpose of this study is to provide a better understanding of communication strategies of the food component in farm tourism with a special emphasis on emotional communication. In fact, this seems to be particularly appropriate for farm tourism operators who want to communicate the oeno-gastronomical "adventure" of the food-and-wine experience (cf. Groß, 2009; Steinecke, 2007).

The following section deals with communication schemes in farm tourism and the relevance of emotional communication for food. We have used practical examples to demonstrate the emotional communication strategies in farm tourism.

3.2 Emotional food communication in farm tourism

According to Gobe (2001), emotions have been a particular influence on the brand management. The consumer buying decision is not only a result of rational decisions. Furthermore, feelings become more important because in many cases consumers are burdened by too much product information. By emotions, Bagozzi et al. (1999) mean "mental states of readiness that arise from cognitive appraisals of events or thoughts" and in commercial terms "are central to the actions of consumers". Concerning food, Alvensleben (2000) underlines in his study that especially for regional food the emotional aspect is very important. In many cases the product quality of regional food cannot be identified by the consumer, so that the buying decision is often a result of emotional attributes of the product or the region in which it is produced.

Most studies which focus on communication in tourism underline the importance of emotions-based communication. The power of oral communication has been examined by Salazar (2005) who traced the linguistic skills that tourist guides develop in order to (re)present and actively (re)construct local culture for a diversified global audience. Chronis (2005) demonstrated how symbols inserted into a text can help to negotiate, define, and strengthen social values between service providers and tourists, whereas Shin, Spark and Moon (2008) showed that the use of metaphors significantly influences the purchasing behaviour of individuals.

Overall, it appears that the more expressive a communication (i.e., the connotative aspects of the message), the more memorable the conveyed information (i.e., the denotative aspects of the message). Against this background, emotional communication seems particularly effective because the connotative aspects such as symbols, metaphors, stories, etc. "reduce the complexity of the reality" (Bosangit, McCabe, and Hibbert, 2009), by providing a more immediate understanding both of the actors involved in the communication process and of the events portrayed.

Although it is plausible that emotional communication can be particularly effective for representing most components of the tourist product, there is less evidence in the tourist literature whether it is appropriate also for conveying information about the food component when the latter is set in the foreground of the tourist experience. In fact, whereas tourism is a dynamic, extra-ordinary and un-contextualized (at least until it is experienced) concept, food is "concrete, contextualized and lived" (Kniazeva and Venkatesh, 2007). This means that it is important to find the right communication for food in tourism since, as underlined by Cohen and Avieli (2004), the culinary aspects of a tourist product can work in its favor or to its detriment:

"Since tourists will be generally reluctant to taste or eat […] foods, whose ingredients are unknown or unfamiliar to them, [the] communication gap between tourist provider and tourists should be reduced in order to avoid any kind of anxiety".

These authors do not prefer any one communication strategy, but they generally affirm that the communication of food in tourism should "filter out" the elements which are unfamiliar to tourists by means of a "cultural translation". The authors refer to the latter as a process which presents the food component anchored to a familiar context. Thus, the communication of culinary aspects in tourism should create an "environmental bubble" (ibid, p. 758) around the tourist. This has a reassuring function which is achieved by using creative techniques such as employing both discursive verbal practices (e.g., labeling dishes with fancy names) or nonverbal ones (e.g., providing pictures of unknown or little known food).

The importance of creative elements in the communication of food in tourism has been confirmed by Charters, Fountain, and Fish (2009). Their work has shown that the provision of a story or a myth can help tourists to connect with the tourist providers in wine tourism. In the same vein, Dehuang, Lixin, and Woodside (2007) have stressed the importance of stories for positioning strategies.

In practice, there are a number of good examples of emotional communication; as in the case of Italy, where agritourism facilities present themselves as a popular venue of regional and vernacular food. Some examples of farm-based accommodation, for instance, use a kind of story with a mixture of romantic, bloody and succulent traits to introduce the facility (www.mutodigallura.com, retrieved October 10, 2009). Others relate their product assortment to familiar subjects like the personification of Italian regional specialties with people such as "Aunt Rosa" (www.ilcanticodellanatura.it, retrieved October 10, 2009) or with poetry (www.laginestra.toscana.it /lingue/EN/prodotti.asp, retrieved January 10, 2010).

These examples not only transmit pertinent information but also provide the reader with more background details, such as the historical development of the farm. This additional information is only partially related to the tangible product as it has, above all, an emotional nature. Hence, it aims less at characterizing the product but at delivering an additional (emotional) value. It is up to the farm operator to decide where to set this emotional value: for instance, the legend which concerns the facility described at www.mutodigallura.com.

Since we think that storytelling could be successfully implemented also by German farm tourism operators to communicate local food, we will devote the next chapter to a detailed description of such a technique.

4 Conceptual framework based on storytelling

Storytelling is a research topic which is discussed in many fields of research, such as economics, literature, psychology, and social science (Brown et al., 2004). Therefore it is hard to find one accepted definition of storytelling. Going back to the basics, storytelling is understood as a narrative form of telling everyday occurrences (Fog et al., 2005).

This way of telling stories can be classified into different literary genres, such as myths or fairy tales. These categories are traditional stories about heroes or other archetypes (Woodside et al., 2008; Groeppel-Klein et al., 2006). The latter can be divided into the *anima* in men and the *animus* in women, representing the hero, and the maiden (Jung, 1982). Jung`s theory of archetypes (1954/59) describes the collective unconscious which can indirectly influence personal feelings and behavior.

Veen (1994) explains that these archetypes have an instinctive or biological function and therefore act as a regulator and stimulator. This means that archetypes activate behavior and therefore play an important role in consumer behavior literature, advertisement and brand management (Groeppel-Klein et al., 2006). Emotional advertisement and brand management that makes use of archetypes is quite common in practice, e.g., "Prinzenrolle" or "Hanuta" in Germany (ibid.) and "Barilla" in Italy.

Furthermore, stories have a typical structure which allows the reader to immediately recognize them as such. These elements create a special story formula. The reader is able to use a script, which is a "predetermined, stereotyped sequence of actions that defines a well-known situation" (Schank and Adelson, 1977) to identify the story (Mangold, 2002). In general, Fog et al. (2005) use four attributes to describe a story: the message, the conflict, the characters and the plot. First of all, a story tries to communicate a message to the reader – a statement that the readers keep in memory. The conflict is "the central turning point of the entire story" (Fog et al., 2005, p. 32) which is often realized in a fight between good and evil, or strong and weak forces. The third feature of a story, the characters, deals with the connection between the hero and the antagonist and their aide(s). All three elements are arranged in the plot creating emotion and excitement.

Based on the theoretical background we define storytelling as an emotional communication of information and experiences by using archetypes and the typical story structure. Even though empirical studies exist concerning consumer behavior and their relation to archetypes, they focus on video or picture analysis (Escalas and Stern, 2003; Gröppel-Klein et al., 2006; Soley, 2006). There are no empirical studies dealing with text analysis using archetypes. A more common methodology in the research is the analysis of responses to commercials concerning different communication styles without integrating archetypes. A study by Adaval and Wyer (1988) showed in three experimental designs that consumers disclose a higher preference for a (holiday) product that is described by means of a common text instead of a simple information list. The question is whether these findings can be transferred on marketing of regional food products. Furthermore Adaval and Wyer (1988) do not examine the effect of a typical story. The aim of this paper is to assess the emotional appraisal of the story in comparison with other (written) communication styles for the promotion of a regional food product. Based on Adeval and Wyer (1988) the authors of this paper suppose that:

> H1: Communication of regional food products based on a story creates the strongest emotional linkage with the reader

Furthermore, the second aim is to detect whether a regional product promoted by a story is more frequently chosen than other similar products that are communicated in different forms, such as a common text or a list of information. The second hypothesis is therefore the following:

> H2: Communication of regional food products based on a story leads to the highest preference for the communicated product.

Based on the above, we have developed an experimental design of three different communication styles which builds on the findings of Adeval and Wyer (1988).

5 Methods

5.1 Data collection and questionnaire design

After a pretest using 60 questionnaires that led to partial modifications of the three different communication styles mentioned above, we interviewed 122 students between June and October 2008. The final number of valid responses was 120. On average, interviewers took about twenty minutes to interview the students, who had been motivated to participate to the survey using incentives.

The questionnaire conceptualized for the survey has the following structure. Overall, there are four main blocks. The first part includes statements which evaluate respondents' attitude towards agriculture in general. The second section contains an experimental design: similar regional food products are presented by three different communication styles (story, text and information list). The third block measures the students' general attitude towards stories and reading activity. Finally, the last section focuses on socio-demographic data of the students. With exception of the last part, the statements are based on seven-point Likert-type scales ranging from "very strongly agree" (+3) to "I do not agree at all" (-3). For the analysis we chose the statistical programme SPSS Version 17. The focus of this paper is mainly on the univariate description of the results.

5.2 Description of the experimental design

As stated above, the experimental design used in our study aims to assess whether story telling has a primary or a complementary role in consumers' stated preference for a regional food product in comparison with other forms of (written) communication.

Therefore, in order to test the hypotheses mentioned above we distinguish among three different communication forms:

(a) A simple information list with product-related data,

(b) A text with the same information but arranged in a narrative form, without any story-related features (Fog et al., 2005) and

(c) A typical story (ibid.) with less information but higher emotional potential.

The writing variations were realized with the support of a scholar of German philology. Additionally, to avoid an influence caused by the order of presenting the texts describing the product, this was varied between questionnaires. Whereas the communication style differs, the background information is always the same and includes the following characteristics:

- The seller is an organic farmer (the name differs in the three texts in order to better adapt to each style),

- The product on sale is pork meat,
- The region where the pig is bred is described in detail.

6 Findings

6.1 Sample description

The mean age of our sample is 25 years. Concerning gender distribution, male respondents are slightly overrepresented (56% versus 44% of female). The academic areas which are most represented are: philology (26%), economics (22.5%) sociology (12%) and agricultural sciences (6%). Overall, respondents in the sample show a positive attitude towards agriculture (mean: 1.45; standard deviation: 1.10) as well as towards organic agriculture (mean: 1.26; standard deviation: 1.4). Furthermore, the questionnaire explores meat consumption habits of the respondents. On average, the students consume meat once or twice a week and most of them consider meat as an important part of the meal (mean: 0.42; standard deviation: 1.9). The high standard deviation scores show, however, a high degree of variance among the answers. The findings concerning the reading habits show that most students usually like to read (mean: 1.99; standard deviation: 2.21). Strictly connected with the above, the ranking of literary genres that are most frequently read shows that novels are the most popular, followed by textbooks. Science-fiction literature is appreciated the least.

6.2 Evaluation of the three communication forms

The following subsection summarizes the main findings of the respondents' evaluation of the three forms of communication administrated during the experiment. The statements discussed in the following are 7-point Likert scaled ranging from totally agree (+3) to totally disagree (-3).

The information list provides useful data summarized into brief items or keywords. Morphologically these are listed by means of different types of bullets (dots or other symbols) which are placed before the text to add emphasis. Since the amount of information provided in the list is very large (size of the agricultural operation, type of certification labels awarded, quality-related issues such as animal welfare and so on) it is not surprising that this communication style does not succeed in entertaining respondents by reading it. Accordingly, the sample mean of the statement "I enjoyed reading it" scores rather low with -0.32 (standard deviation: 1.74).

The text differs from the list of information in many ways. Morphologically, items are not listed anymore. Moreover, from the point of view of the syntax, the text consists of grammatically correct sentences which are subsequently bound together (text coherence). Apparently, these few differences between the list of information and the text are sufficient to "mislead" respondents who to the statement "I am reading a story" score the list of information very low (mean: -2.2; standard deviation: 1.02), thus acknowledging the information list as such, but show a higher level of uncertainty about the text (mean: 0.66; standard deviation: 1.63). Hence, a larger share of respondents is persuaded to read a story instead of a text. Finally, the high mean scores to the statement "I enjoy reading it" (mean: 0.97; standard deviation:

1.36) show that this form of communication is overall well appreciated since it elicits fun while reading.

As mentioned above, in the field of literature and literary criticism the story is a well-defined literary genre of its own (Hudson, 1960; Jensen, 1999; McKee, 1997). Indeed, it is recognised as such by the majority of respondents (mean: 2.2; standard deviation: 0.98). Moreover, the story scores high with regard to the statement "I enjoy reading it" (mean: 0.97; standard deviation: 1.36) although to a lesser extent than the text.

Table 1 displays the salient traits of the three writing forms for three main statements: the image of the farmer, the emotional appraisal and the interest in the communicated product. Among the three writing forms the text provides the better image of the farmer who stands "behind" the product. In fact, to the statement "This writing form provides a clear image of the farmer" the text scores the highest (mean: 1.84; standard deviation: 1.30), followed by the story (mean: 1.16; standard deviation: 1.84) and last by the information list (mean: -0.48; standard deviation: 2.02). As mentioned, the second row of the table focuses on the emotional linkage of the communication style with the reader. The highest degree of emotion is delivered by the story (mean: 0.47; standard deviation: 1.60) followed by the text (mean: -0.50; standard deviation: 1.70). Hence, since the communication style based on the story displays the highest emotional appraisal on respondents we can accept the first hypothesis of our study.

Further, we want to detect whether a high emotional linkage with the readers leads also to a higher purchasing intention. A preliminary analysis of this is displayed by the last row of Table 1. Indeed, we notice that there are differences among the mean scores for the statement related with the interest in the product communicated by each writing form. The respondents of our sample state that the information list does awake interest in the product (mean: 1.00; standard deviation 1.69) although, as previously shown, this form of communication is not considered particularly entertaining. The communication form which follows immediately after is the text (mean: 0.97; standard deviation: 1.36). Surprisingly, the story delivers a negative mean score (mean: -0.79; standard deviation: 1.67).

Table 1. Evaluation of the three forms of communication

This [form of communication] elicits...[1]:	Information list	Text	Story
... a clear image of the farmer	-0.48 (2.02)	1.84 (1.30)	1.16 (1.84)
... emotions	-2.4 (0.9)	-0.5 (1.7)	0.47 (1.6)
... interest in the product	1.00 (1.69)	0.97 (1.36)	-0.79 (1.67)
[1]Seven-point Likert-type Scale: from -3 = totally disagree to 3 = totally agree			

Source: own study

The low impact of the story in terms of product interest in individuals is further confirmed by Figure 1. The percentage of respondents who somewhat to very strongly agree to the statement "I would buy the product described in the text" is

75.8%, followed by 72.5% who somewhat to very strongly agree to the statement "I would buy the product described in the information list."

Figure 1. Purchase intention

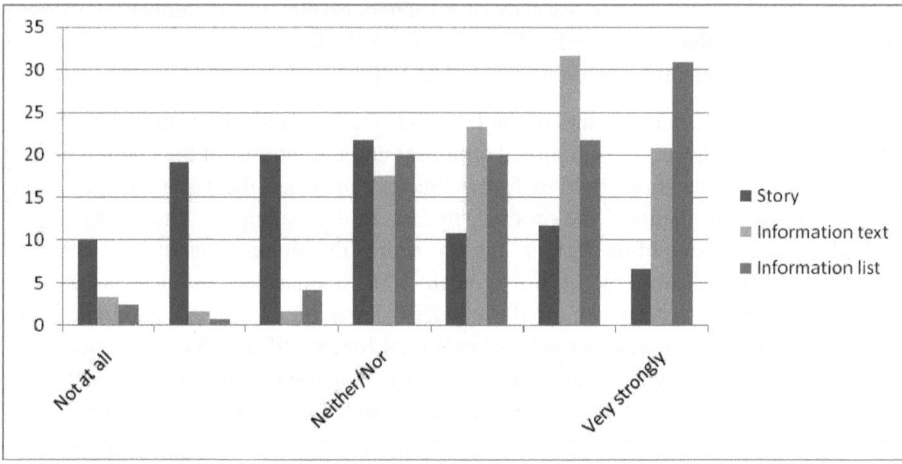

Source: own study; "I would buy the products from this farmer" Scale: from (-3) = I do not agree at all to (+3) = I very strongly agree

Finally, only 30% of respondents disclose a somewhat to very strong agreement to the statement that they "would buy the product communicated by the story". Because of the low share of 30% of participants attracted by the story we cannot accept hypothesis 2.

7 Conclusions and limitations

This article deals with the appropriate communication strategies for the promotion of regional food products in combination with farm tourism, as this is an important aspect of the farm tourism supply in several European countries such as Italy and Austria. Since also in Germany farm tourism operators are increasingly trying to distinguish themselves as providers of high-quality food, the purpose of the study was to explore whether emotional communication strategies (i.e., story-telling) are appropriate to communicate such product types.

Although in the tourist literature emotional communication strategies are considered a driving force for the product preference, our study does not empirically support evidence of it. Only 30% of our sample seems to be positively influenced by a product description conveyed by a story as they would by the product. Hence, all in all, it seems that a very emotional communication technique, as story-telling is, is not the appropriate strategy to influence customers' preferences for regional product.

A first conclusion that we can draw is that the relationship between communication strategies and type of communicated product may be much closer than generally assumed. Emotional communication, as shown by many studies, is suitable for the overall promotion of tourist products since these belong to the so called "experience economics" and as such, their power of attraction for customers consists of the novelty and excitement aspects they transmit to potential tourists. So, for instance, farm tourism is often promoted by features which are far away from the day-to-day experience such as the *leitmotiv* of adventure (dining by a campfire; Boskovic, 2003; Williams et al., 2001), romanticism (mystique of rural areas; Wilson et al., 2001), introspective travel (tourist as the ethnographer of the self, Ingram, 2002; Quan and Wang, 2004), etc. It is evident, that such features are successfully transmitted through emotional communication strategies.

However, when we switch to the 'material culture' of food and beverages, the values that have to be communicated are different. Food habits of individuals are notoriously entrenched and on holiday this resistance to exogenous changes may increase as shown by Cohen and Avieli (2004). Therefore, as it is shown by our findings, it is plausible to assume that communication of (regional) foods should exploit well-known rather than little known characteristics, reassuring rather than innovative features.

The information conveyed through the text succeeds in influencing the buying intention of respondents more successfully than the story and the information list. It seems that this form of communication is appropriate to provide the "environmental bubble" stressed by Cohen and Avieli (2004) in order to convey the traditional and authentic characteristics of regional food without depriving the individual from the familiar context as, probably, the strategy of story telling does. It is legitimate to assume that the values of authenticity and tradition, which are of paramount importance for regional food (Stockebrand and Spiller, 2008) are preserved by the text but seem to be lost by the story and, to a lesser extent, by the information list.

Before concluding, we should point out the study's limitations. Being a pilot study with an explorative approach a convenience sample was used. Furthermore, the experiment depends on the linguistic style. Nevertheless, the study offers some directions for future research development. Farm tourism has been studied in several countries with different approaches. Yet at present, there are no studies that have investigated communication strategies of farm tourism operators who want to distinguish themselves as providers of high-quality catering.

Based on the findings of the study, it would be interesting to analyze the attitude towards emotional communication strategies on the internet or in conjunction with other imagery-based tools like pictures, videos, etc. Thus, the potential of German farm tourism especially in combination with culinary heritage has not been fully exploited yet. Further studies in these directions could highlight new opportunities for this tourist sector.

References

Adams, B. (2008). *The new agritourism. Hosting community and tourists on your farm*. Auburn, State of California: New World Publishing.

Adaval, R. and Wyer, R. (1998). The Role of Narratives in Consumer Information Processing. *Journal of Consumer Psychology*, *7*(3) 207-245.

Albersmeier, F., Schulze, H., and Spiller, A. (2007). System Dynamics in Quality Certifications: Development of an Audit Quality Controlling System. *International Journal on Food System Dynamics 1*(1), 69-81.

Alvensleben v., R. (2000). Zur Bedeutung von Emotionen bei der Bildung von Präferenzen für regionale Produkte. *Agrarwirtschaft*, 49, 399-402.

ARM (Azienda Romana Mercati) (Eds.) (2007). *Voglio aprire un agriturismo*. Roma: Agra.

BAG (Bundesarbeitsgemeinschaft für Urlaub auf dem Bauernhof / German association of farm tourism) (Ed.) (2008). *Activity report 2007/2008.*

BAG (Bundesarbeitsgemeinschaft für Urlaub auf dem Bauernhof / German association of farm tourism) (Ed.) (2009). *Activity report 2008/2009.*

Bagozzi, R. P., Gopinath, M., and Nyer, P.U. (1999). Special issue on the past, present, and future of the marketing discipline - The role of emotions in marketing. *Journal of the Academy of Marketing Science, 27*(2), 184-206.

BMELV (Bundesministerium für Ernährung, Landwirtschaft und Verbraucherschutz / German Ministry for food, agriculture and consumer protection) (Ed.) (2006). *Urlaub auf dem Bauernhof / Urlaub auf dem Lande*. Studie für das Bundesministerium für Ernährung, Landwirtschaft und Verbraucherschutz auf der Basis der Reiseanalyse 2006 (Study for the BMELV on the basis of travel analysis 2006). Bonn.

BMELV (Bundesministerium für Ernährung, Landwirtschaft und Verbraucherschutz / German Ministry for food, agriculture and consumer protection) (Ed.), (2008). *Expertise "Urlaub auf dem Bauernhof / Lande"* . Bonn.

Bosangit, C., McCabe, S., and Hibbert, S. (2009). What is told in travel blogs? Exploring travel blogs for consumer narrative analysis. In W. Höpken (Ed.), Information and communication technologies in tourism 2009. *Proceedings of the international conference in Amsterdam*, The Netherlands, 2009, 517. Vienna: Springer.

Boskovic, D. (2003). *Education for the development of village tourism in Montenegro*. Republic of Montenegro: Ministry of Education and Science. Retrieved June 10, 2009, from http://www.gov.me/files/ 1084171165.pdf.

Brown, J., Broderick, A. J., and Lee, N. (2007). Word of mouth communication within online communities: Conceptualizing the online social network. *Journal of Interactive Marketing, 21*, 2-20.

Brunori, G., Di Iacovo, F., Marescotti, A., Pieroni, P., and Rossi, A. (2003). Comunicazione e sviluppo rurale: considerazioni preliminari per un'agenda di ricerca. *Rivista di economia agraria, LVIII*(2), 271-294.

Charters, S., Fountain, J., and Fish, N. (2009). "You Felt Like Lingering.": Experiencing "Real" Service at the Winery Tasting Room. *Journal of travel research, 48*(1), 122-134.

Chronis, A. (2005). Coconstructing heritage at the gettysburg storyscape. *Annals of Tourism Research, 32*(2), 386-406.

Clarke, J. (1996). Farm accommodation and the communication mix. *Tourism management, 17*(8), 611-616.

Cohen, E. and Avieli, N. (2004). Food in tourism - Attraction and Impediment. *Annals of Tourism Research, 31*(4), 755-778.

Dehuang, N., Lixin, C., and Woodside, A. G. (2009). Storytelling research of consumers' self-reports of urban tourism experiences in China. *Journal of business research, 62*(12), 1223-1254.

Eghbal, D. (2008). Tourismus als Standbein. *Bioland Fachmagazin für den ökologischen Landbau 1*(2008), 21-22.

Embacher, H. (1994). Marketing for agritourism in Austria: strategy and realization in a highly developed tourist destination. *Journal of tourism research, 2*(2), 61-76.

Escalas, J. E. and Stern, B. B. (2003). Sympathy and empathy: emotional responses to advertising dramas. *Journal of Consumer Research, 29*, 566-578.
Gobe, M. (2001). *Emotional branding: the new paradigm for connecting brands to people*. New York: Allworth Press.
Fog, K., Budtz, C., and Yakaboylu, B. (2005). Storytelling. Branding in Practice, Berlin: Springer.
Groß, M. S. (2009). Tourismus und Gastronomie im Wellness-Trend. In A. Dreyer (Ed.), Schriftenreihe Dienstleistungsmanagement: *Tourismus, Sport, Kultur* (Vol. 2009, 245-262). Hamburg: ITD-Verlag.
Gröppel-Klein, A., Domke, A., and Bartmann, B. (2006). Pretty woman or Erin Brockovich? Unconscious and conscious reactions to commercials and movies shapes by fairy tale archetypes-results from two experimental studies. *Advances in Consumer Research, 33*, 163-174.
Hashimoto, A., and Telfer, D. J. (2006). Selling Canadian culinary tourism: branding the global and the regional product. Tourism Geographies: *An international journal of tourism space, place and environment, 8*(1), 31-55.
Hudson, W. H. (1960). *An introduction to the study of literature*. London: Harrap.
Ingram, G. (2002). Motivations of farm tourism hosts and guests in the South West Tapestry region, Western Australia: A phenomenological study. *Indo-Pacific journal of phenomenology 2*(1), 1-12.
Jensen, R. (1999). *The dream society: How the coming shift from information to imagination will transform your business*, New York: Mc Graw-Hill.
Jung, C. G. (1954/1959a). The archetypes and the collective unconscious. In H. Read, M., Fordham, and G., Adler (Eds.), *The collected works of C. G. Jung* (pp. 3-41). Princeton, NY: Bollingen.
Jung, C. G. (1982). *Aspects of the feminine*. London: Routledge.
Kniazeva, M., and Venkatesh, A. (2007). Food for thought: a study of food consumption in postmodern US culture. *Journal of consumer behaviour, 6*(6), 419-435.
Kuhl, I. R. and Niessen, I. (2009). Ausrichtung der Marketingstrategien deutscher Bio-Handels- und –Herstellerunternehmer auf die LOHAS-Zielgruppe. Paper presented at the *10. Wissenschaftstagung Ökologischer Landbau*. Mayer et al. (Eds.). Retrieved September 5, 2009, from http://orgprints.org/14056/1/ Kuhl_14056.pdf.
Mangold, M. (2002). Markenmanagement durch Storytelling. München: Fördergesellschaft Marketing.
McKee, R. (1997). *Story: substance, structure, style, and the principles of sceenwriting*. New York: Harper-Collins Publishers.
Murray, I. (2008). Culinary tourism: segment or figment? Unpublished work. Retrieved June 8, 2009, from http://people.okanagan.bc.ca/arice/2008TTRA Proceedings/2008%20 Proceedings/Papers/Murray,%20Iain.%20%20Culinary%20Tourism%20-%20Segment% 20or%20 Figment.pdf.
Murray, I.P., and Haraldsdottir, L. (2005). Developing a rural culinary tourism product: considerations and resources forsSuccess. ICTA Conference, San Francisco, June 2005.
Nilsson, A. P. (2002). Staying on farms - An ideological background. *Annals of tourism research, 29*(1), 7-24.
Oekolandbau in Deutschland (2008). Retrieved June 6, 2008, from: www.bmelv.de/cln_044/nn_750590/SharedDocs/downloads/04-Landwirtschaft/ Oeko Landbau/ Oekolandbau__in__ Deutschland__ Tabelle2, templateId =raw,property =publication File.pdf/ Oekolandbau _in _Deutschland _Tabelle2.pdf.
Oppermann, M. (1995). Holidays on the farm: a case study of German hosts and guests. *Journal of travel research, 34*(1), 63-67.
Oppermann, M. (1996). Rural tourism in Southern Germany. *Annals of tourism research, 23*(1), 86-102.
Przezbórska, L. (2003). Relationships between rural tourism and agrarian restructuring in a transitional economy: The case of Poland. In Hall, D., Mitchell, M., and Roberts, L. (Eds.), *New directions in rural tourism* (pp. 203-222). Aldershot: Ashgate.

Quan, S., and Wang, N. (2004). Towards a structural model of the tourist experience: an illustration from food experiences in tourism. *Tourism management, 25*(3), 297-305.

Salazar, N. B. (2005). Tourism and glocalization "local" tour guiding. Annals of Tourism Research, 32(3), 628-646.

Schank. R. C., and Abelson, R. P. (1977). *Scripts, plans, goals, and understanding*, Hillsdale, NJ: Psychology Press.

Schöppner, A. (1988). Urlaub auf dem Bauernhof. Eine fremdenverkehrsgeographische Untersuchung der Angebots- und Nutzerstrukturen in der Bundesrepublik Deutschland. Bochum: Geograph. Inst., Ruhr-Univ.

Schulte-Sasse, J., and Werner, R. (2001). Einführung in die Literaturwissenschaft. München: Fink.

Sharpley, R., and Vass, A. (2006). Tourism, farming and diversification: An attitudinal study. *Tourism management, 27*(5), 1040-1052.

Shin, J. K., Park, M. S., and Moon, M. K. (2008). The effects of metaphors in advertisement – comparison between experience goods and search goods. Paper presented at the Global Marketing Conference, Shanghai, March, 20-23, 2008.

Soley, L. (2006). Measuring reponses to commercials: A projektive-elicitation approach. *Journal of current issues and research advertisment,* 28(2), 55-64.

Spiller, A., Voss, J., and Deimel, M. (2007). Das EU-System zum Schutz geographischer Herkunftsangaben und Ursprungsbezeichnungen: Eine vergleichende Studie zur Effektivität des Instruments zur Förderung des ländlichen Raums und Implikationen für die deutsche Agrarförderung. In Rentenbank (Eds.), Wettbewerbsfähigkeit der deutschen Agrarwirtschaft – politische, institutionelle und betriebliche Herausforderungen. Schriftenreihe der Landwirtschaftlichen Rentenbank, 22, 187-232.

Steinecke, A. (2007). *Kulturtourismus - Marktstrukturen, Fallstudien.* München/Wien: Perspektiven.

Stockebrand, N., and Spiller, A. (2008). Authentizität als Erfolgsfaktor in der Regionalvermarktung – eine erste Skizze. In Antoni-Komar, I., et al. (Eds.): *Ernährung, Lebensqualität – Wege regionaler Nachhaltigkeit* (pp. 145-166). Marburg: Metropolis.

Telfer, D. J., and Wall, G. (1996). Linkages between tourism and food production. *Annals of tourism research, 23*(3), 635-653.

Veen, S. V. (1994). The consumption of heroes and the hero hierarchy of effects. Advances in Consumer Research, 21, 332-336.

Wagner, P., Burger, H.-G., and Magnus, R. (1997). Ferien auf dem Bauernhof: Was wünschen die Urlauber? Ergebnisse einer Befragung. Frankfurt am Main: DLG-Publ.

Williams, P., Paridaen, M., Dossa, K., and Dumais, M. (Centre for Tourism Policy and Research) (2001). Agritourism Market and Product Development Status Report, BC: British Columbia Ministry of Agriculture, Food and Fisheries.

Wilson, S., Fesenmaier, D. R., Fesenmaier, J., and van Es, J. C. (2001). Factors for success in rural tourism development. *Journal of Travel Research, 40*(2), 132-138.

Wohlmann, R., and Lohmann, M. (1986). Urlaub auf dem Bauernhof, Urlaub auf dem Land. Eine empirische Untersuchung der Meinungen, Einstellungen und Verhaltensweisen von Urlaubern. Starnberg: StfT.

Woodside, A. G., Sood, S., and Miller, K. E. (2008). When consumers and brands talk: storytelling theory and research in psychology and marketing. *Psychology and Marketing, 25*(2), 97-145.

Policy options for sustainability. a preliminary appraisal of rural tourism in Romania: the case of Maramureş

Francesca Regoli
Matteo Vittuari
Andrea Segrè

Department of Agricultural Economics and Engineering
Faculty of Agriculture
University of Bologna, Italy
{francesca.regoli, matteo.vittuari, andrea.segre}@unibo.it

Abstract

The shift of agriculture from the sole production of food and fibre to the provision of services has led to the development of a new rural development paradigm focused on diversification, rural non-farm economics, local resources, and social interactions. Among diversification strategies, rural tourism (RT) has emerged as one of the main drivers of change and major contributors to the sustainable exploitation of local resources. In this framework, the article aims to investigate the role and potential of RT in the county of Maramureş (Romania) focusing on the connection of the offer of rural accommodation (RA), the territory, and the strategies for the valorisation of local cultural and natural resources (human capital and territorial respectively). The first part of the research work is based on an extensive literature review and analysis of the secondary data to identify the key characteristics of the region and the role of RT within the diversification strategies adopted in farming and rural systems. The second part focuses on the RA offer within farm activities. The cluster analysis is conducted to group the farms according to the quality level of the bid for RA and to investigate the connection of the offer with local resources.

Zusammenfassung

Durch die Erweiterung klassischer landwirtschaftlicher Aktivitäten der Nahrungsmittel- und Faserproduktion hin zum Angebot von Dienstleistungen entstand ein neues Paradigma der ländlichen Entwicklung, das auf Diversifikation und außerlandwirtschaftliche Einkommen sowie auf lokale Ressourcen und soziale Interaktionen fokussiert ist. Unter den Diversifikationsstrategien ist der Landtourismus einer der wichtigsten Treiber des Wandels, der maßgeblich zur nachhaltigen Nutzung lokaler Ressourcen beiträgt. Vor diesem Hintergrund ist es das Ziel dieses Artikels, die Bedeutung und das Potenzial des Landtourismus in der Region Maramures (Rumänien) zu analysieren. Im Mittelpunkt steht dabei die Verbindung zwischen dem Angebot von Unterkünften auf dem Land, der Region und den Strategien zur Wertschöpfung aus lokalen kulturellen und natürlichen Ressourcen (regionsspezifisches sowie Humankapital). Der erste Teil des Beitrags identifiziert, basierend auf einem umfassenden Literaturüberblick sowie der Auswertung von Sekundärstatistiken, die wesentlichen Charakteristika der Region und die Bedeutung des Landtourismus unter den Diversifikationsstrategien der landwirtschaftlichen Betriebe ebenso wie des ländlichen Systems. Der zweite Teil widmet sich dem Angebot von Unterkünften als Teil der Aktivitäten landwirtschaftlicher Betriebe: Basierend auf der angebotenen Qualität der Unterkünfte und der

K.L. Sidali et al. (eds.), *Food, Agri-Culture and Tourism: Linking local gastronomy and rural tourism: interdisciplinary perspectives*, DOI 10.1007/978-3-642-11361-1_3,

jeweils vorhandenen lokalen Ressourcen werden landwirtschaftliche Betriebe mittels einer Clusteranalyse zu Gruppen zusammengefasst.

Riassunto

Il passaggio da un'agricoltura basata sulla mera produzione di cibo e fibre ad un'attività estesa alla produzione di servizi, ha portato ad un nuovo paradigma dello sviluppo rurale interessato alla diversificazione, ad un'economia rurale non prettamente agricola, alle risorse locali ed alle interazioni sociali. Tra le strategie di diversificazione, il turismo rurale (TR) emerge come uno dei principali fattori di cambiamento e di utilizzo sostenibile delle risorse locali. In questo contesto, l'articolo si sofferma sull'analisi del ruolo e delle potenzialità del TR nella regione del Maramureş (Romania) concentrandosi sul rapporto tra l'offerta di alloggi da parte delle aziende agricole, il territorio e le strategie di valorizzazione delle risorse locali, culturali e naturali (capitale territoriale ed umano). La prima parte del lavoro di ricerca analizza lo stato dell'arte e le fonti secondarie sulle principali caratteristiche dell'area e sul ruolo del TR nell'ambito di strategie di diversificazione adottate nei sistemi agricoli e rurali. La seconda parte esamina l'offerta di alloggi da parte delle aziende agricole tramite un'analisi cluster: le aziende agricole sono state raggruppate in base al livello di qualità dell'offerta ed alla sua relazione con le risorse locali.

Keywords: Maramureş; Sustainable development policy; Rural accommodation; Human capital.

1 Introduction

Romania has a significant tourist potential, which includes seaside tourism (Black Sea), skiing and hiking (Carpathian mountains), cultural heritage tourism (Bucovina and Transylvania), rural tourism (Maramureş and Bucovina), spa and wellness tourism (Black Sea and Maramureş), and conference and exhibition tourism (Bucharest).

Within this framework and with consideration of the predominance of rural regions in Romania, rural tourism (RT) represents a major driver for the promotion of the economic vitality of the countryside and achievement of the goals identified in the National Development Strategy 2013-2020-2030. According to this document, a set of objectives are developed, including improvement of environmental quality in rural areas, enhancement of the quality of life in countryside, support for diversification of rural economy, strengthening of the competitiveness in agricultural, forestry, and fisheries sectors, and promotion of local development initiatives. To meet these objectives, policies have to guarantee the maintenance of a social balance connecting quality of life to economic development, as requested by the Axis 3 of the European Rural Development Policy 2007-2013. Therefore, the diversification of rural economy is as a major policy aim for the sustainable exploitation of local resources and RT as a major driver of change.

Human resources and capacity building processes play a key role in this framework. The potential of the tourism sector and its impact on the development of rural areas include the valorisation of the local human capital in connection to the natural capital (Brunori, 2006). The article focuses on the Maramureş county with the aim to analyse the role of RT within the diversification strategies adopted in farming and rural systems. Through the use of a cluster analysis, the article investigates the quality level of the bid for rural accommodation (RA). Finally, the connection between the offer and the local resources is assessed.

2 Theoretical background

2.1 Sustainable tourism within the new rural development paradigm

The multifunctional role of agriculture has been widely recognized and investigated by specialized literature. Previous studies explore diverse agricultural issues from the production of food and fibres to the provision of services such as soil conservation, water and land management, agro-ecological system conservation (Van Huylenbröck et al., 2006), biodiversity protection, and energy production. In this framework, a new "actor oriented" rural development paradigm has emerged (OECD, 2006) with local actors considered as drivers of change (Murdoch et al., 2003; Brunori, 2007) and with area as the product of social space shaped through different social interactions reflecting various human practices (Mahon, 2007). The wider attention on actors implies a stronger emphasis on the process of establishing partnerships among different local stakeholders around a common objective (i.e., the valorisation of the territory and its capital), as well as bringing outside actors (tourists, consumers, citizens) within the projects implemented by the local community (Brunori, 2007).

This new paradigm was embodied in the European policies drafted by European Councils held in Lisbon (2000) and Göteborg (2001). The four Axes of the Rural Development Policy 2007-2013 emphasise the competitiveness of the agricultural and forestry sectors (Axis 1), environment and countryside (Axis 2), quality of life in rural areas and diversification (Axis 3), local capacity for employment and diversification (Axis 4 – Leader). These Axes led towards the sustainable management of natural resources, increased investments in human capital, promoted a more balanced access to productive assets, markets and services, and created more participatory and accountable institutions (Commission of the European Communities, 2002).

Therefore, the adoption of a quality approach for sustainable management of the resources is a primary aspect (Dubost, 2008; Bruckmeier and Tovey, 2008) in the planning and implementation of policies and projects with the involvement of local knowledge (Kloppenburg, 1991; Tovey, 2008). For this reason, the creation of a set of indicators to measure the improvement of local actions and policies towards a sustainable approach represents an opportunity to strengthen the link between scientific and policy discourses (Bruckmeier and Tovey, 2008). Moreover, an additional relevant instrument is represented by the certification schemes for tourism quality: a tool to improve performance of tourist operators and increase awareness about sustainability both in the tourist offer and on the demand side (UNEP 2006).

Different international organizations (UNWTO, 2003; UNEP 2006; EU through the Ecolabel[1] scheme) emphasise the role of training, communication, and networking activities in the promotion of certification schemes with the purpose of increasing capability and quality level of the workforce employed in the tourist sector as well as enhancing awareness of more sustainable and responsible procedures. UNWTO

---

[1] http://www.ecolabel-tourism.eu

(2003) asserts certification schemes for sustainable tourism should be designed according to the geographical, political, and socio-economic features of a country. The special attention is paid to the "integral component" of the sustainable tourism development strategy.

2.2 Sustainability issues in rural tourism

The specialized literature emphasises the diverse contributions and linkages that tourism creates with other sectors of the economy (OECD, 2010). It is also the case that rural tourism extends its connections with a number of community stakeholders. Such interactions appear evident in planning and developmental schemes chosen by governments aimed at increasing integration of economic, social, and environmental objectives. The result is the recognition of challenges faced by the tourism sector including globalization processes, climate change and global warming, environmental and social sustainability, impacts on labour markets and human resources, good governance, and innovation (OECD, 2010).

This increased attention towards an integrated approach to tourism has led to a shift from conventional mass tourism to alternative forms of tourism characterized by elevated attention to the effective use of environmental resources, respect of the socio-cultural authenticity of the host communities, and promotion of long-term economic strategies to ensure a fair distribution of benefits among all the stakeholders (UNWTO, 2004). Meanwhile, positive externalities are also generated by natural resources at the local and global level as producers of services: i.e., water filtration and purification provided by wetlands, or carbon sequestration that helps in mitigating global warming, provided by forests (OECD, 2008). Overall sustainability of tourism, including rural tourism, is largely based upon a proper exploitation of the territorial capital and local natural, human, and social resources.

Table 1. Mass tourism vs. rural/eco-tourism

	Intensive mass tourism	**Rural/eco-tourism**
Local cultural resources	High commercialization of handcrafts and traditional heritage	Valorisation of traditional heritage
Local natural resources	High pressure on local resources as water, energy, food and other raw materials	Creation of governance systems for the preservation of local resources
Land degradation	High or increasing level of anthropisation	Low flows of tourists, low level of anthropisation, protection of the wilderness
Air pollution and noise	Massive use of air, road and rail transport systems	Promotion of alternative low impact systems as walking, trekking, biking
Waste	Increasing amount of garbage	Resource efficiency
Physical impacts	Mass infrastructures, large hotels, large commercial areas	Traditional accommodations and family guesthouses
Biodiversity	Alteration of ecosystems and natural habitats	Attention paid to ecosystems
Alternative employment	Large number of seasonal employers	Prevalence of local employers
Financial contributions	Prevalence of international investors and corporations	Prevalence of family run businesses

Source: authors' elaboration

3 Methodological issues

The first part of the paper consists of an extensive literature review and provided the analysis of secondary data to identify the main characteristics of the region, the role of RT within the diversification strategies adopted in farming and rural systems, and to evaluate the instruments and programmes used to promote the Maramureş as a tourism destination. Collected data was utilized to analyze specific sustainability issues and to investigate local options and policy choices aimed at the development of rural areas.

The second part of the work is dedicated to the analysis of farms offering tourist hospitality in Maramureş county. Information on accommodation (198 structures) and services was gathered from electronic sources including websites of specific RT networks, local administrations and communities, individual websites of rural accommodation, and emails. Moreover, the electronically collected information was integrated with a number of interviews carried out with operators and local experts.

A numerical value was assigned to each attribute to process the collected data through a cluster analysis, following the *Un-weighted Pair-Group Method using arithmetic Average* (UPGMA, Romesburg, 2004). The classification allowed identification of different groups among the guesthouses according to the qualitative level of the rural accommodations' offer and their link with territorial capital. The algorithm, based on the UPGMA approach, was created using the Matlab® programming language[2]. A resemblance matrix was created to identify the similarities among the objects. The resemblance matrix was turned into a graphic tree to identify the clusters. Each cluster grouped objects with similar characteristics.

4 Towards a Romanian rural development policy

4.1 The Romanian rural development policy behind agriculture

Following the European Rural Development Policy 2007-2013, the Romanian National Rural Development Programme (NRDP) focuses on increasing the competitiveness of the agricultural sector, improving quality of life in rural areas and ensuring diversification of the rural economy. As emphasised by the National Strategy for Romanian Sustainable Development, tourism plays a significant role in the accomplishment of these goals due to the high interest of Romanian Government in the sector (Ministry of Environment and Sustainable Development, 2008).

---

[2] http://www.mathworks.com

To this end, rural areas in Romania, following the national definition[3], cover 87.1% of the territory and engage 45.1% of the population or 9.7 million of inhabitants (National Statistical Institute - INS, 2005). Agricultural land covers 62% of the total, and the country has a strong rural character. Romania has the highest share of rural population (45.1%) among the European countries. However the situation is significantly different at the regional level, as most municipalities have fewer than 50 inhabitants/km^2 in the Western part. In the Eastern and Southern parts, municipalities have a density of between 50 and 100 inhabitants/km^2 (NRDP, 2009).

As in many other transition countries, Romanian rural areas are characterised by a prevalence of elderly people and a high share of the active population engaged in the agricultural sector. In Romania, these trends are particularly consistent since one third of the rural inhabitants is occupied in agricultural activities (32%) and about the 34% of the rural actives are over 55 years. Youths are abandoning rural areas and agricultural sector due to lack of alternative employment opportunities and low level of services, which have a significant impact on the overall quality of life. In general, the rural-urban migration and migration to foreign countries are extremely relevant issues. Estimates reveal that over 2 million Romanians (about 10% of the total population) are working abroad in Germany, Greece, Italy and Spain (MAPDR, 2009).

Another barrier for the improvement of farmers' economic stability and living standards is represented by the current farm structure and utilized agricultural area (UAA). About 80% of the UAA is shared by 80% of small farms with less than 5 ha and a small number of farmers with over 50 ha operating 40% ha of the UAA. The remaining 20% of UAA is shared among farms with an average size from 5 to 50 ha (NRDP, 2009).

Moreover farmer associations are exceptionally weak both on production and marketing side (National Rural Development Programme 2007-2013). The small farm dimensions and lack of efficient professional organisations represent a limitation for local farmers in contracting opportunities. To address this challenge, the National Development Strategy 2013-2020-2030 recommends the introduction of a bottom-up approach for governance of rural communities. The strategy emphasises the role of Local Action Groups (LAGs) in the creation of networks among local stakeholders to implement projects and initiatives at the community level (Axis 4). In this evolving framework, RT represents one of the main tools (Van Huylenbroeck, 2006) within farm diversification strategies, especially considering the cross-sectoral nature of tourism (Grenna et al., 2006) and the number of potentially interested stakeholders.

---

[3] The territory is administratively organized in 319 municipalities and cities for the urban area and 2,851 communes for rural areas, according to the 2nd Level of the Local Administrative Units -LAU (former NUTS 5 level). As stated by the Law 351/2001, about territorial planning and urbanism, communes are mostly made up of more than one village (12,946 villages in total which do not have any administrative responsibilities). Romanian Ministry of Agriculture, Forestry and Rural Development - MAPDR (2008), *Raportul Annual de Progrese privind implementarea programului național de dezvoltare rurală în România în anul 2007,* June.

Nevertheless, in the Romanian countryside, the rural non-farm activities remain largely unexplored. In 2005, non-farm activities represented 4.1% (approximately 12 Euros/month) of the average net income of rural households (MAPDR, 2009). To revitalise the rural economy in the non-agricultural sector and allow increase of incomes, both the improvement of skills through training (i.e., entrepreneurial and business administration) and adoption of a different approach towards the agro-rural world (generally not perceived by the population as a centre of innovation and development) are fundamental elements which require full attention.

The current situation in the RT sector illustrates difficulties in exploiting rural resources other than the agricultural ones. The current tourism trends show a limited average length of stay as the number of foreign tourists hosted in accommodation structures represents only 22.2% of the total capacity (Romanian INS, 2007). The challenge is partially related to the quality standards of accommodation (absence of a national quality standard influences the overall quality of national tourism offers), lack of supplementary recreational activities, i.e., the promotion of tourist itineraries to raise awareness about the country tourism offers at a national and international level, and absence of working government strategy (World Travels & Tourism Council - WTTC, 2005). In addition to these weaknesses, the Romanian National Institute of Statistics recorded increasing in flows in the tourist sector: the arrivals in tourist facilities moved from 5,000 in 2002 to more than 6,200 in 2006 (INS, 2007). The tourism sector represents a potential driver of change for a sustainable growth and valorisation of the country; nevertheless, considering the experiences of the central and eastern European countries that joined EU in 2004, a "*project class*" is expected as an essential requisite for the successful design, implementation, and administration of the initiatives.

4.2 Rural tourism as diversification strategy in Maramureş

Shifting from the national to the regional level, one of the most interesting cases is represented by the county of Maramureş. The district covers 2.6% of the national territory and hosts numerous villages preserving the traditional rural culture and heritage. Small communes (ref. to Law 351/2001) are the most widespread form of agglomeration. In fact, the county is composed of 2 municipalities, 11 cities, and 63 communes (Council for Maramureş district, 2009). Agricultural land covers 11.6% (74,524 ha) of the total with the agricultural population representing, in official statistics, more than the 14% of the active population. However, agriculture represents an important safety net for a wider share of rural population which is still involved in subsistence or semi-subsistence farming (Salasan et al., 2009). Despite the high percentage of people involved in agriculture, the small dimension of farms (subsistence and semi-subsistence farm households), age structure, and limited numbers of farmers with an adequate technical and economical knowledge reduce the spread of modernization and diversification strategies, which could offer additional economic and social opportunities considering the territorial capital of Maramureş.

Due to lack of entrepreneurial capacities, the economic potential of the area is largely unexplored. Local resources are more often spoiled than valorised (OECD Rural

Policy reviews, 2006). Considering the tourist sector, weak entrepreneurial capacity is highlighted by inadequate tourist infrastructures. It appears that the hosting facilities in Maramureş are poor both in terms of number and quality. Data show that only 17.6% of the total potential capacity is operating in Maramureş (INS, 2007). The most common typology of accommodation in the region is represented by the rural tourist boarding houses (87), followed by hotels and motels (24) and urban tourist boarding houses. According to the national law (L. 638/08), these facilities are classified as rural accommodation (*pensiuni agroturistice*) that are evaluated through *daisies* and other guesthouses evaluated with stars. However, the classification in daisies and stars is rather unreliable due to ineffective certification process (assessed by law every 3 years). Therefore, the information provided by tourist and non-institutional websites is significantly different from the one appearing in official statistics. In the case of the Maramureş county, the difference between official figures on rural accommodations and the terrain situation ranges from 87 to 198. Rural accommodation are classified according to eleven attributes aimed at describing a mix of qualitative and territorial elements. The parameters selected by the authors are mainly based on the national legislation. Law 638/08 defines structures with farm activities or connected to the production of local handicraft as rural accommodation. The law indicates that rural accommodation should respect a maximum of 8 rooms in independent buildings or in dedicated spaces of the house, preparation of meals from local food (directly from the farm or from local producers), and offer possibility for guests to participate in farm and other local activities. The national legislation framework is combined with the criteria fixed by the Italian law on agritourism (Law 96/2006), Eurogites (Product definition for "Rural Accommodation", General Assembly, 2005), and a number of Italian local institutions operating in the tourism sector (i.e., Municipality of Faenza, Trade Chamber of Crotone). According to the indicated framework, the following parameters are taken into consideration as qualitative elements:

- capacity of the accommodation units: number of rooms on a scale from 1 to 8 (as identified by the national law 638/08);
- compliance with the national standards: number of daisies attributed on a scale from 1 to 5 (national law 638/08);
- services included in a room on the basis of a classification on three categories: high standards (private bathroom, private TV, internet access, air conditioned), medium standards (air conditioned, private TV), and low standard (none of the previous requirements);
- promotion of rural accommodation and related activities through a web site: presence of a individual web-site;
- capacity to assist foreign tourists in foreign languages: number of spoken foreign languages on a scale from 1 to 5.

Qualitative attributes were further combined with territorial elements considering the following:

- location of rural accommodation: in the open countryside or within a village;
- morphology of the area where rural accommodation is located: mountain (above 600 m), hilly (from 200 to 599 m), or flat area (from 0 to 199 m);

- overall assessment of the authenticity of accommodation and its ambience: respect for the traditional architecture, coherence of internal decorations, and access to the local culture and traditional heritage;
- capacity to offer traditional local foods: availability of gastronomic offer;
- overall assessment of the offer of complementary activities available at local level (attractions, monuments, natural sites, excursions, and sport activities): high level of connection (3 to 4 activities), medium level (1 to 2 activities), and low level (0 activities);
- involvement of women in management of rural accommodations (as referent person): gender inclusiveness.

According to the numerical values assigned to the above attributes, a resemblance matrix (a tree) was elaborated. One tree covered all attributes and the other two trees connected to the qualitative and territorial features. For the quality tree, 10 clusters were taken into consideration. Cluster 2 and 6 appeared the most representative including 33 and 146 guesthouses respectively. The other clusters, grouping a total of 8 rural accommodations, were defined as "strange fallen objects" (Romesburg, 2004) since they included few rural accommodations and, therefore, were not significant for the analysis.

Table 2. Main characteristics of clusters 2 and 6

Cluster 2 (33 rural accommodations)	Cluster 6 (146 rural accommodations)
More than half of the guesthouses have no web-site	Almost no guesthouse has a web-site
Most of the hosts speak one foreign language	Most of the hosts speak almost one foreign language
The number of rooms is between 6 and 7	The number of rooms is between 3 and 4
Most of the guesthouses have 2 daisies (the average appears higher than in cluster 6)	Most of the guesthouses have 2 daisies
Offered services are at a medium/high standard	Offered services are at a medium standard

Source: author's elaboration

Most of the analyzed structures offered a medium standard service, characterised by a rather small size without any attention paid to promotional activities (i.e., creation of a website). Lack of promotional strategy reduced the diffusion of information to potential tourists through the online promotion organised by regional networks (i.e.,

ANTREC, MTMM[4]), which offer more standardised information on a number of guesthouses.

Together with an appropriate promotion, the capacity to provide information in foreign languages is a significant driver to attract foreign tourists and increase the confidence to visit a country which is still scarcely known outside its own borders. For the territorial tree, 6 clusters were considered.

Table 3. Main characteristics of clusters 6, 1, 5 and 4 in order of representativeness

Cluster 6 (75 rural accommodations)	**Cluster 1 (42 rural accommodations)**
Equally located in the countryside and outside villages	Almost half of the accommodations are located outside the village
Located in hilly areas	The majority of guesthouses are located in hilly areas
Characterized by an authentic ambience	Most of the accommodations have a gastronomic offer
A medium/high connection with the territory	The link with the territory is at a medium level
Cluster 5 (39 rural accommodations)	**Cluster 4 (31 rural accommodations)**
Most of the accommodations are located within the village	Almost all accommodations are located within the village
Almost all the guesthouses are in hilly areas	Almost half of the guesthouses are in hilly areas
Almost half of the accommodations are characterized by authenticity of the ambience	Almost all of the accommodations are characterized by authenticity of the ambience
Managed just by women	Managed mainly by men or it was not indicated
Most of the accommodations have no gastronomic offer	All the accommodations have no gastronomic offer
The link with the territory is weak	The link with the territory is weak

Source: authors' elaboration

Four clusters appeared the most representative from a total of six: clusters 1, 4, 5, and 6, grouping respectively 42, 31, 39, and 75 RA. Cluster 6 is the most related to the

[4] ANTREC - National Association of Rural, Ecological and Cultural Tourism; MTMM - Land of Maramureş Microregion Association, OVR – Opération Villages Roumaines: http://www.antrec.ro/;http://www.mtmm.ro/index.php?jump=agroturism;http://www.ovr.ro /Frameset%20Romanian.htmlhttp://www.visitmaramures.ro/;www.ruraltourism.ro;www.kazare .ro.

territory. RAs are trying to offer different kinds of activities connected to the region such as organized itineraries (visits to typical wooden churches, excursions with traditional carts, free entrance or discount to spa centres or ski facilities, and visits to ethno museums) and tasting experiences. Cluster 1 differs from the previous ones mainly for the management (gender inclusiveness is rather low) and a more limited connection to the territory. Cluster 4 is located in village areas managed by men. The guesthouses are characteristic, even if the connection to the territory is present through a gastronomic offer or complementary activities, is exceptionally weak. Cluster 5 is rather different from the other clusters as the management is by women, but RA does not appear authentic, there is no gastronomic offer, and lack of significant connection to the territory.

5 Conclusions

Romania has a strong rural character with a large share of population engaged in or dependant on the agricultural sector. Considering rural development policies, the diversification of farm activities was introduced only with the *National Rural Development Programme 2007-2013,* which is still incompletely implemented. The multifunctional dimension of farm activities remains widely unexplored among rural households, and local administrations do not make significant actions to foster the modernisation process. This framework reflects the situation in Maramureş county. Despite this, multifunctionality and rural non-farm economics represent a unique opportunity for the development of the Romanian countryside. The article aimed to analyse a specific element related to multifunctional agriculture: rural tourism and its features, development, and potential in the Maramureş region. The analysis focused on those farms offering rural accommodation (according to the national law 638/08). There was a discrepancy in the total number registered between official statistics and data collected through electronic sources and classified according to the criteria set in the national law (attribution of daisies).

As mentioned in the article, the overall awareness of the local population of the potential offered by the tourism sector to integrate their income is limited. Many managers among the analysed accommodations have not fully explored the potential of their capital. The value of the rural tourism experience is based on the quality of the *countryside capital*, including social, natural, and physical capital (Garrod, 2006), which need to be redefined.

The cluster analysis allowed the identification of the major groups and emphasised their recurrent elements and distinctive characteristics. The most representative clusters are generally characterised with a medium level standard of services and weak territorial network. The emerging scenario showed a limited networking capacity of rural guesthouses, while considering the prevalence of small communes in the Maramureş county, the collaboration among these small entrepreneurs represents an important asset for a wider coordination and representativeness towards regional and national administration institutions. The creation of a regional body dedicated to tourism management[5] could represent an important driver to increase quality standards and create an effective territorial network. Local human, social, and natural resources should be combined and coordinated to create a unique and distinct product to be valorised in a specific promotional plan (Mettenpenningen et al., 2009). Nowadays, tourist destinations are becoming more competitive (Buhalis, 1999); therefore, being innovative, competent, and strongly interconnected to the territory represents a strategic tool to guarantee stability to Maramureş district. Local, regional, and national development strategies should promote economic growth with consideration of social and environmental issues to foster balance between sustainability and competitiveness (Dubost, 2008)[6].

Acknowledgements

The creation of the algorithm (through the Matlab® programming language) has been possible thanks to the collaboration of the colleague Dr. Filippo Zaniboni, Department of Physics – Geophysics Section, University of Bologna. The research has been supported by the Vigoni programme (promoted by the Italian-German University DIH and the German Institution DAAD) aimed at favouring scientific exchanges among Italian and German researchers. The exchanges, in particular, took place between the research group lead by Prof. A. Spiller, GA Göttingen University, and the research group lead by Prof. A. Segrè, University of Bologna.

[5] The creation of a regional organization has been foreseen in the National Strategic Plan of Rural Development 2007-2013.

[6] Further research will be conducted to analyze the weight of the hosting activity within the total income of farmers in order to explore the overall relevance of rural tourism and to identify a number of good practices.

References

Bruckmeier, K., and Tovey, H. (2008), Knowledge in sustainable rural development: from forms of knowledge to knowledge processes, *Sociologia Ruralis*, Vol 48, Number 3, July, Blackwell Publishing, 313-329

Brunori, G. (2006), Post-rural processes in wealthy rural areas: hybrid networks and symbolic capital, in *Between Local and the Global: confronting complexity in the contemporary agri-food sector,* Research in "Rural Sociology and Development" (Vol. 12, 125-152), Elsevier Ltd.

Brunori, G., and Rossi A. (2007), Differentiating countryside: social representations and governance pattern in rural areas with high social density: the case of Chianti, Italy, *Journal of Rural Studies 23*, 183-205.

Buhalis; D. (1999), Marketing the competitive destination of the future, *Tourism Management 21* (2000), 97-116.

Camera di Commercio Crotone, Agriturismi di Qualità, Criteri di valutazione per assegnare il marchio di qualità negli agriturismi.

Commission of the European Communities (2002), Communication from the Commission *Fighting Rural Poverty* – COM (2002) 429 final, 25.7.2002, Brussels.

Comunità Montana dell'Appennino Faentino, Servizio associato per le attività produttive, Modello D1 – Modello di Autocertificazione. Richiesta di assegnazione del livello di classifica ai sensi della L.R. 26/1994 art. 32 e D.G.R. n. 389/2000.

Council for Maramureş District (2006), Document de Programare Dezvoltarea Turismului în Judeţul Maramureş 2007-13, Draft 4, October.

Dubost; M. (2008), Risultati ed effetti concreti riguardo al miglioramento competitività e della sostenibilità, *Leader + Magazine 11*, 43-49.

European Commission (2007), LEADER+ type activities in ROMANIA, *Country Sheet - Member State Files*, September 2007.

European Commission, Agriculture Directorate-General: Rural Development, European Communities (2006), *Fact Sheet - New Perspectives for EU Rural Development,* European Communities.

European Commission, Directorate-General for Agriculture and Rural Development (2008), *The EU Rural Development Policy: Facing the Challenges,* European Communities, September.

Garrod, B., Wornell R., and Youell R. (2006), Re-conceptualising rural resources as countryside capital: The case of rural tourism, *Journal of Rural Studies* 22, 117-128.

Government of Romania - Ministry of Agriculture, Forests and Rural Development (2008), "Context and Legal framework for the implementation of the Leader Axis in Romania".

Government of Romania - Ministry of Agriculture, Forests and Rural Development (*2009), National Rural Development Programme 2007-2013 – consolidated version 31 March 2009*, Number CCI: 2007RO06RPO001.

Government of Romania - Ministry of Environment and Sustainable Development (2008), *Strategia Năţională pentru Dezvoltare Durabilă a României Orizonturi 2013-2020-2030,* Programul Naţiunilor Unite pentru Dezvoltare Centrul Naţional pentru Dezvoltare Durabilă, Bucureşti.

Grenna, L. (World Bank); Hilbruner R. (USAID); Santi E. (World Bank); Scuppa G. (World Bank); Vereczi G. (UNWTO) (2006), Communication and Sustainable Tourism, *Proceedings of the Global E-Conference and Summer Speaker Series on the role of Development Communication in Sustainable Tourism* May 29 - June 9, 2006.

Kloppenburg, J. Jr (1991), Social theory and the de/reconstruction of agricultural science: local knowledge for an alternative agriculture, *Rural Sociology* 56 (4), 519–548.

Mahon, M. (2007), New populations; shifting expectations: the changing experience of Irish space and place, *Journal of Rural Studies 23*, 345-356:

Monitorul Oficial nr. 408/24 July 2001, L. Nr. 351/2001 privind aprobarea Planului de amenajare a teritoriului naţional - Secţiunea a IV-a reţeaua de localităţi.

Murdoch J., Lowe, P., Ward, N., and Marsden, T. (2003), *The differentiated countryside,* London: Routledge.

National Institute of Statistics - Maramureş District Direction of Statistics in http://www.maramures.insse.ro/main.php?lang=fr&pageid=42C.

National Institute of Statistics (2007), *Statistical yearbook 2007 - Anuarul Statistic al României - Tourism 20*(2007).

OECD (2006), *The new rural paradigm, policy and governance,* Paris: OECD Publishing.

OECD (2008), *Natural Resources and pro-Poor Growth. The Economics and Politics,* OECD Publishing.

OECD (2010), *OECD Tourism Trends and Policies 2010,* Paris: OECD Publishing.

Romesburg, H. C. (2004), *Cluster Analysis for Researchers,* North California: Lulu Press.

Salasan, C., and Fritzsch, J. (2009), The role of farm activities for overcoming rural poverty in Romania. In Burchenrieder, G, and J. Möllers (Eds.) *Structural Change in Europe's Rural Regions – Farm Livelihoods Between Subsistence Orientation, Modernization and Non-farm Diversification*, Vol. 49 of IAMO Studies Series (IAAE 2009 Mini-symposium), 23-41.

Tovey, H. (2008), Introduction: rural sustainable development in the knowledge society era, *Sociologia Ruralis*, *48*(3), 185-199.

United Nations Environment Programme (2006), Tourism certification as a sustainability tool: assessment and prospects, January, United Nations Environment Programme.

UNWTO (2003), *Recommendations to governments for supporting and/or establishing national certification system for sustainable tourism*, WTO-STD e-bulletin, World Tourism Organization, Madrid.

UNWTO (2004), *Sustainable Tourism Resource Management*, World Tourism Organization, Spain.

Van Huylenbroeck, G. (2006), Multifunctionality or the role of agriculture in the rural future, *The Rural Citizen Governance, Culture and Wellbeing in the 21st Century*, Compilation © 2006 University of Plymouth, UK.

World Travels & Tourism Council (2006), *Romania the Impact of Travel & Tourism on Jobs and the Economy*, World Travels & Tourism Council, London.

From agricultural to rural: agritourism as a productive option

Rossella Bianchi

Department of Engineering and Management of Agricultural, Livestock and Forestry Systems - PRO.GE.S.A.
University of Bari Aldo Moro
rossella.bianchi@agr.uniba.it

Abstract

This paper intends to focus on the transition from a "productivist" and farm business development model - dominant until the 1970s within the framework of the Common Agricultural Policy (CAP) - to a multifunctional, environmentally sustainable and integrated paradigm in which the primary sector opens up to become rural, acquiring a systemic, diversified, alternative, and locally sustainable productive perspective. In the 1980s, this model evolved by means of a fundamental, and principally cultural, development. It has shaped rural development policies, marking the down-sizing of old price and market policy. First of all, the intention is to sketch a brief evolutionary-historical outline of the CAP orientation; agritourism is then identified as a product innovation *à la Schumpeter* and an interesting diversification of farm productive activities. Agritourism is analysed from a multifunctional perspective, aimed at safeguarding the integrity of natural resources and also at enhancing socio-cultural aspects and the oeno-gastronomical heritage of specific rural areas. There is also a brief review of agritourism, present in the Italian juridical system from the 1980s, seen from the perspective of a researcher into economic theory: attention is focused specifically on Law No. 96/2006.

Zusammenfassung

Dieser Beitrag fokussiert auf den Übergang von dem in der Gemeinsamen Agrarpolitik der EU in den 1970er Jahren vorherrschenden produktivitätsorientierten Modell der landwirtschaftlichen Betriebsentwicklung zu einem Paradigma der multifunktionalen, ökologisch nachhaltigen Landwirtschaft. Der Primärsektor wird hierbei als Teil des ländlichen Raums mit einem systemischen, diversifizierten und lokal nachhaltigen Fokus konzipiert. Die so veränderte ländliche Entwicklungspolitik ist durch die Rückführung der klassischen Instrumente der Markt- und Preispolitik gekennzeichnet. Im Folgenden soll zunächst ein kurzer Überblick über die Entwicklung der Ausrichtung der GAP im Zeitablauf gegeben werden; daraufhin wird der Agrartourismus als eine *Schumpeterianische* Produktinnovation und interessante Diversifikationsstrategie für landwirtschaftliche Betriebe beschrieben. Dabei wird der Landtourismus in multifunktionaler Hinsicht analysiert, der zur Erhaltung natürlicher Ressourcen ebenso wie zur Stärkung soziokultureller Aspekte und des önogastronomischen Erbes spezifischer ländlicher Regionen beiträgt. Darüber hinaus wird ein kurzer Überblick über den Agrartourismus, der seit den 1980er Jahren im italienischen Rechtssystem verankert ist, aus Sicht der ökonomischen Theorie gegeben. Der spezielle Fokus liegt dabei auf dem Gesetz Nr. 96/2006.

K.L. Sidali et al. (eds.), *Food, Agri-Culture and Tourism: Linking local gastronomy and rural tourism: interdisciplinary perspectives*, DOI 10.1007/978-3-642-11361-1_4,

Riassunto

Nel presente lavoro, relativamente all'orientamento delle politiche agricole comunitarie, si intende focalizzare il passaggio da un paradigma di sviluppo di stampo aziendalistico e produttivistico, prevalso fino agli anni '70, ad uno di tipo integrato, multifunzionale ed ecosostenibile, in cui la dimensione del primario si apre a quella del rurale, acquisendo una prospettiva produttiva sistemica, diversificata, alternativa e localmente sostenibile. Quest'ultimo paradigma si è via via definito nel corso degli anni '80 attraverso una fondamentale evoluzione di carattere innanzitutto culturale. Essa ha sedimentato la sua logica nelle più recenti politiche di sviluppo rurale, segnando il ridimensionamento della vecchia politica dei prezzi e dei mercati. Nel tracciare un sintetico profilo storico-evolutivo di tali politiche, ci si propone di inquadrarvi l'agriturismo quale *schumpeteriana* innovazione di prodotto, interessante forma di diversificazione delle attività produttive aziendali tradizionalmente intese, inquadrata in una prospettiva pienamente multifunzionale, volta a tutelare l'integrità delle risorse naturali nonché valorizzare gli aspetti socio-culturali ed eno-gastronomici delle specifiche realtà rurali. Il valore dell'agriturismo è colto inoltre, sin dagli anni '80, nelle scelte normative dell'ordinamento giuridico italiano, nel quale ci si propone di compiere una breve incursione, pur con gli occhi di chi studia la realtà dal punto di vista dell'evoluzione economica, soffermandosi sul significato delle scelte compiute dalla L. n. 96/2006.

Keywords: Rural development; Agritourism; Product innovation.

1 Introduction

This paper intends to focus on the transition from a "productivist" and farm business development model - dominant until the 1970s within the framework of the Common Agricultural Policy (CAP) - to a multifunctional, environmentally sustainable and integrated paradigm in which the primary sector opens up to become rural, acquiring a systemic, diversified, alternative, and locally sustainable productive perspective. In the 1980s, this model evolved by means of a fundamental development which is principally cultural. It has shaped rural development policies, marking the down-sizing of old price and market policy.

First of all, the intention is to sketch a brief evolutionary-historical outline of the CAP orientation; agritourism is then identified as a product innovation *à la Schumpeter* and an interesting diversification of farm productive activities. Agritourism is analysed from a multifunctional perspective, aimed at safeguarding the integrity of natural resources and also at enhancing socio-cultural aspects and the oeno-gastronomical heritage of specific rural areas. There will also be a brief review of agritourism, present in the Italian legal system from the 1980s, seen from the perspective of a researcher into economic theory.

2 Productivity and sectorality in the CAP of the 1970s

At the end of the 1970s, the prevailing CAP model shows clear signs of crisis. The nearly exclusive implementation of the price and market policy has bolstered increasingly intensive productive processes in the agricultural systems of the Member States, with widespread use of chemicals and a high level of mechanisation (von Meyer, 1983; Priebe, 1985); concentration of production activities in the flat, well-watered areas with a very great potential for economic and productive development

(Fabiani, 1986). The consequent imbalances have involved, at the microeconomic level, the allocation of inputs, productive patterns, and disadvantages for small and medium farms (De Benedictis, 1981); and, on the wider system level, environmental imbalances, distortions in international trade, and the accumulation of huge food surpluses (von Meyer, 1983). Policies of export paybacks, price integration and guaranteed prices above the world market level have created a situation of unsustainable "irrational subsidy", -"die subventionierte Unvernunft" (Priebe, 1985). Attempts by the (then) EEC in the 1970s to harmonize different development levels among the various agricultural areas were not successful, despite the three directives in 1972 and the directive for mountainous and less favoured areas (EEC Directive No.268/1975). These directives express a new structural policy, but still encapsulate a farm and sector-based concept of agricultural development (Fanfani, 1996; Mantino and Pesce, 1996).

3 International pressures and constraints on the CAP in the 1980s

In 1987, the OECD (Organisation for Economic Co-operation and Development) publishes a report commissioned in 1983, whose findings are very important. The Producer Subsidy Equivalent, estimated for countries with a mature capitalist economy on data for a three-year period '79-'83, is on average more than 30%; little more than 40% for the EEC; more than 50% for Japan and North Europe countries. This highlights a huge amount of public expenditure to support agricultural prices and incomes. Moreover, this expenditure is unevenly allocated since only 25% of farmers profit from 75% of this amount (OECD, 1987). Evidently, the problem does not involve only the European CAP, and during the GATT negotiations, in the framework of the Uruguay Round, there is some very strong pressure to liberalize international trade and so reduce agricultural prices to the world market levels.

The political and international context has changed and so has the balance of power. The process of market internationalization which began in the late 1960s has now developed fully (Graziani, 1998) and the perspective of a new globalized economy is emerging. New developing countries try to enter into this new context and they are ready to trade more freely (Josling, 1993). Furthermore, countries that were formerly excluded from western capitalism are now beginning to take part in commercial dynamics. In these years, the system of multilateral world trade changes very quickly. So, a new approach to international trade and problems requires an innovative set of rules (Josling, 1993). The issue reaches to a crucial point regarding the agro-food product sector because it is affected by a wide range of both physiological and pathological impacts, caused by agricultural policies at the EEC and national levels.

In 1983, there is the Stuttgart Summit, and in March 1984 an agreement is reached on agricultural prices. The two events mark a definite U-turn in the direction of the CAP (Saccomandi, 1988). An EEC summit is held in Brussels in February 1988 which leads to a crucial decision: reduction of the average price level for each agricultural product within the EEC Common Market Organization. It is a question of reducing EEC agricultural expenditure and giving a new balance to the price and market policy, restoring priority to the variable "price" in its meaning of orientation in the productive choices of farmers.

4 The transition from agricultural to rural in the theoretical model of the 1980s

So the time is ripe to bring a new degree of sensitivity and attention into the direction of Common Agricultural Policies, going beyond considerations strictly concerned with production. Official documents and Regulations approved by the EEC in the 1980s are an effective expression of the transition from agricultural to rural which takes place in the theoretical, political and cultural model. It is interesting to re-read the text of the new structural regulations: the regulation on "improving efficiency of agricultural structures" (EEC Reg. No. 797/85); the regulation about "agricultural structures and adjustment of agriculture to the new market situation, and to the preservation of the countryside" (EEC Reg. No. 1760/87); the regulation concerning "setting aside arable land and the extensification and conversion of production" (EEC Reg. No. 1094/88), which modified the two previous regulations.

Although the lines traced in terms of converting arable land, extensification and set aside land aim to address the problems of re-establishing the market equilibrium and limiting the amount of surplus production, they actually introduce innovative issues of environmental protection, countryside preservation and support for agriculture in less-favoured areas, with social and cultural aims of protecting the landscape and its resources.

The differentiated specific regional problems are recognized, and a new rural development policy is approved for the less-favoured areas, emblematically indicated by the integrated systemic approach contained in the IMPs, (Integrated Mediterranean Programmes, EEC Reg. No. 2088/85); a lucid analysis is made of the CAP and of the problems it caused, and of the possibilities for re-establishing equilibrium within the options given in Andriessen's Green Paper of July 1985; unprecedented sensitivity towards economic and social cohesion is affirmed with the adoption of the Single European Act (EEC, 1987); and above all there is a new interpretation of the problems of the primary sector, which are carefully differentiated and analysed according to specific territorial features, contained in the fundamentally important theoretical and programming document: "The Future of Rural Society" (CEE, 1988).

5 CAP and multifunctionality

The political and cultural synthesis of the 1980s sees increasingly deeper consideration of the environmental and rural aspects of regional development, highlighting a new sensitivity towards the problems involved. These aspects begin to influence economic issues and development processes, and to support the orientation of the reforms planned by the EEC. This process is evident from the Reform of the Structural Funds in 1988 (EEC Reg. No. 2052/88) with its integrated and programmed vision for rural development cycles renewed at five-yearly intervals up until 2007-2013 and still ongoing; to the Mac Sharry Reform in 1992 with its principal of compensation in relation to a sharp reduction in prices, and the supplementary package of agricultural and environmental regulations (EEC Reg. No. 2078/92 and No. 2080/92), up to Agenda 2000 and the Fischler Reform in 2003 (EC Reg. No. 1782/2003) which definitively downsized the old CAP scheme with its principles of decoupling, modulation and cross-compliance in order to promote rural development, environmental quality, food security, a more competitive market oriented agriculture, more efficient administrative procedures.

In the context of the present policies for the primary sector, the framework used for analysis and interpretation of development problems is that of multifunctionality. This means that agriculture can no longer limit its activities to the production of surpluses of qualitatively undifferentiated food products, but now has to provide quality foods, organoleptically intact and that are typical and specific according to the territorial area of origin, the production techniques used and the adequate connections in the production chains. At the same time, agriculture also has to protect the landscape, safeguard the natural resources and biodiversity, enhance cultural and oeno-gastronomical traditions of the region and produce positive externalities on the environment and for society as a whole (Bianchi, 2010).

From this viewpoint, new productive dimensions are developing in the rural context, alternative to the traditional farm and sectoral models. Parks are increasingly important, as are nature reserves, nature paths, pony trekking, agritourism or rural tourism, as it has been called since the 1980s in EEC documents and regulations (EEC Reg. No. 214/84). These alternatives have created space for culture and education, leisure and, more widely, for the diffuse interests of the entire society, for physical and mental re-integration in an environmental context where the integrity of natural resources is protected, as well as the specific architectural, historical and cultural features of the rural areas (Bianchi, 2001). In this way, even agro-food products acquire a series of intangible meanings linked to geographical specificity, and to the value given to the typical production methods belonging to the history of one place and not repeatable elsewhere (Brunori, 2003).

6 Diversification: a survival strategy

In this way, a different interpretation of the problems of the primary sector has emerged, together with the development of a wide variety of functions of European agricultures which are not obviously connected exclusively to production. The CAP also recognizes this in the Health Check, and can therefore justify to the WTO payments to farmers which are not connected with production quantities (De Filippis, 2008). This marks the decline of the sectoral development model based on undifferentiated support for agriculture (Henke and Salvioni, 2008).

In the long term, European farmers are increasingly oppressed by reduced agricultural incomes caused by the marked divergence between sale prices and production costs. They may react to the difficulties of this situation with product innovations *à la Schumpeter*, and by introducing new combinations of production factors, meeting new segments of market demand, diversifying production and reducing costs.

Stimulated by multifunctionality, the new directions of EU policy (Reg. EC No. 1698/2005), and their own survival instinct, farmers develop diversification strategies and try out innovative activities. These include countryside management, conservation of natural resources, agritourism, organic farming, biodynamic farming, and the production of high quality and specifically local products using low environmental impact farming methods, and also product processing on the farm, direct sales, educational farms, and social and therapeutic agricultural activities (Mantino, 2008; van der Ploeg et al., 2000).

In this way, farmers, in an innovative entrepreneural *habitus,* try to find a place in the niches provided by a rural economy which is integrated, diversified, multifunctional and environmentally, economically and socially sustainable.

7 Agritourism: a product innovation

Agritourism plays an emblematic role in the passage from agricultural to rural, or alternatively from a model of an intensive and sectoral agriculture to a model of territorial development which is endogenous, multifunctional and locally sustainable. It may be interesting, at the theoretical level to take another look at Schumpeter's analysis in his Theory of Economic Development (Schumpeter, 1971 and 2002).

According to Schumpeter, innovation and development in an economic system are generated by all the processes which involve a new combination of production factors. This happens in the following cases:

1. Production of new goods
2. Introduction of new production methods
3. A new market opening up
4. Identification of a new source of raw or semi-processed materials
5. Re-organization of any production sector.

Agritourism is definitely a new means of combining farm production factors and - in a situation of multifunctional diversification - is itself an innovative product and a new type of supply in the primary sector. The dimension of this supply opens up a new market and directs consumer demand for the material and immaterial goods and services which multifunctional agriculture can provide and to which reference has already been made.

According to neo-classical theory, the consumer is a central priority and attributes economic value to consumer goods, expressing his marginal utility towards them. His needs and preferences generate production and the production itself aims at giving him the opportunity to repeat his activity of consumption. According to Schumpeter, this is the case for static routines, but in innovative situations or those which can generate development of the system, development begins with the entrepreneur who proposes a new product.

Therefore Schumpeter defines a theoretical perspective, an alternative to the dominant cultural model of his age, placing the innovative entrepreneur at the centre of his theory of economic development, and beyond the well-tried production routines. With the strength of a Prometheus, the innovative entrepreneur goes against the mainstream and implements innovations on the free market, not only in terms of technology, in the process of "creative destruction" which is a feature of endogenous development within the capitalist system.

Although Schumpeter's analysis is extremely interesting, it is really necessary to bear in mind the specific features of his age, if we now want to carry out a new in-depth analysis of economic development and draw from it some useful categories for analysis and interpretation of the present context. Schumpeter's work was produced at a time when state intervention in the economy was not a consolidated fact, neither at the level of institutional theory nor at the level of political and administrative practice.

Nowadays, despite the crisis in the social market economies connected also with the decline of a Ford-type production system (Rullani, 1997), the political and regulatory system of the EU and the institutional and juridical systems of its Member States are still able to propose development options and production stimuli to orientate specific strategies in entrepreneurial behaviour. This is what happens, as seen at the level of promoting forms of production diversification and entrepreneurial directions, like agritourism, which are evidently linked not only to the attitudes of the economic subject but also to the stimuli given by the political, institutional and juridical context.

To this regard, one of the axes of the rural development programme 2007-2013 in the most recent EU policy is the fundamental objective of diversification of the rural economy and of improving the quality of life in rural areas (EC Reg. No. 1698/2005). However, if we look back to the past, rural tourism is first mentioned in a 1984 Regulation aimed at promoting - especially in disadvantaged agricultural areas - "a specific community action to contribute towards the development of certain French and Italian regions in the context of enlargement of the Community"[1] (Reg. EEC No. 214/84); and Andriessen's 1985 Green Paper which began to introduce a regional perspective for analysis of the problems of re-establishing social, economic and environmental equilibrium. Among other options, here there was an explicit reference to encouragement of rural tourism as a form of diversification of farmers' incomes. These were falling due to the introduction of eco-compatible farming practices aimed at reducing the amount of surplus production which was no longer possible to sustain in financial and environmental terms.

So it was EC (European Commission) choices which sent out the first signals to stimulate new combinations of production factors, creating renewed spaces for interaction between supply and demand for goods and services, and between the processes of accumulation and redistribution of resources (Bianchi, 2004).

[1] The enlargement regarded Greece in 1981; Spain and Portugal in 1985.

For this reason, using Schumpeter's categories, we could assert that new combinations of production factors are achieved also through process innovations (Schumpeter, 1971 and 2002) which are exquisitely immaterial (Botta, 2001) regarding, for example, the political and legal framework in which the developmental problems of a society are interpreted and tackled (Bianchi, 2004).

8 Agritourism and process innovation

The present Italian legal framework governing "Regulation of Agritourism" is Law No. 96/2006 which has its roots in the fundamental 1980s text (Law No. 730/1985).

Table 1. Regional laws regulating agritourism based on Law February 20, 2006, No. 6

Italian regions and A.P.[1]	Regional laws (LR)	Title
Valle d'Aosta	LR December 27, 2006, No. 29	New agritourism regulations
Piedmont	LR March 23, 1995, No. 38	Agritourism regulations
Liguria	LR November 21, 2007, No. 37	Regulations for agritourism, fishing and fish tourism
Lombardy	LR June 8, 2007, No. 10	Regional agritourism regulations
A.P. Trento	LP December 19, 2001, No. 10	Regulations for agritourism, and wine and food routes
A.P. Bolzano	LP September 10, 2008, No. 7	Agritourism regulations
Veneto	LR April 18, 1997, No. 9	New regulations for the agritourism industry
Friuli Venezia Giulia	LR July 22, 1996, No. 25 LR November 9, 1998, No. 13 LR February 22, 2000, No. 2 LR October 17, 2007, No. 25	Agritourism regulations Modifications and integrations Modifications and integrations Modifications and integrations
Emilia Romagna	LR March 31, 2009, No.4	Regulations for agritourism and the multifunctionality of farms
Marches	LR April 3, 2002, No. 3	Regulations for the agritourism industry and rural tourism
Umbria	LR August 14, 1997, No. 28 LR August 12, 1989, No. 31	Regulations for the agritourism industry Modifications and integrations
Lazio	LR November 2, 2006, No. 14	Regulations for agritourism and rural tourism
Abruzzo	LR May 31,1994, No. 32 LR March 4, 1998, No. 12 LR February 24, 2003, No. 4	New regulations for agritourism Modifications and integrations Modifications and integrations
Molise	LR June 16, 2001, No. 13	Interventions in favour of agritourism and the development of rural areas
Campania	LR November 6, 2008, No. 15	Regulations for the agritourism industry
Apulia	LR May 22, 1985, No. 34	Interventions in favour of agritourism
Basilicata	LR February 25, 2005, No. 17	Agritourism and rural tourism
Calabria	LR April 30, 2009, No. 14	New regulations for agritourism, educational and social activity on farms
Sicily	LR June 9, 1994, No. 25	Regulations for agritourism
Sardinia	LR June 23, 1998, No. 18	New regulations for agritourism

Source: retrieved October 18, 2009, from http:// www.agriturist.it [1]A.P.: autonomous provinces

This is the national legal framework; the Italian regions establish their regulations within the lines of this framework (see Table 1).

The 2006 law is wide-ranging and systemic; its aims fully recognize the multifunctional nature of agritourism in relation to support for agriculture, and to increasing and diversifying agricultural incomes (Art. 1, Law No. 96/2006). It tends to consider agritourism as a vital component of the local system, able to activate positive dynamics for the integration of agriculture with the economy and society (Fabiani, 1991), with positive effects on the regional economy in terms of territorial competitiveness (Brunori, 2003), on the cohesion of social networks (Di Iacovo, 2003) and on the local ecological and environmental equilibrium (Van der Ploeg, 2006; Mantino, 2008).

This is shown by the schemes of processing, valorisation and marketing of regional, typical and local agro-food products, not only from the farm itself, which agritourism operators are obliged to use in their catering (Art. 2, para. III from *b.* to *d.* and para. IV; Art. 4, para. IV from *a.* to *d.*). It is also interesting to consider direct selling by farmers, which is a way of internalizing the stages of the food production chain within the farm, with positive externalities on the environment at the same time. This sales-strategy avoids the environmental pollution caused by long-distance transport and by the often complex commercial packaging necessary for large-scale retail distribution (Henke and Salvioni, 2008). There is also the positive and revitalizing social effect of *empowerment* of rural communities deriving from their direct contact with the tourist-consumer (Henke and Salvioni, 2008).

Of course, both the physical-geographical context and the social economic and cultural context of the agritourism farm are fundamentally important. This means the position of rural areas near to urban centres with efficient infrastructures, and with a governance able to implement policies to stimulate and support the development processes undertaken; the existence of technical and productive abilities and embedded codified expertise, essential for valorising the locally typical oeno-gastronomical heritage; and the availability of architectural, artistic, historical and cultural features which can offer visitors for interesting and alternative tourism and leisure activities. These leisure or cultural activities may take place independently of accommodation and catering activities, as long as they are connected "to the agricultural activity and resources of the farm", and to other activities to encouraging exploration of the local historical, environmental and cultural heritage of the territory (Art. 4, para. V).

The law also projects agritourism as part of a multifunctional approach for the recovery and valorisation of the rural architectural heritage, and later regional regulations cover *ad hoc* interventions for this purpose (Art. 3, para. II). This is an opportune formula for environmental resilience, because it provides for the conversion and renovation of existent rural buildings, for example disused stalls or barns which are converted into farm shops. The aim is to convert buildings which contribute to the quality of the environment, and respect the specific features and architecture of the regional area, as well as the landscape and surroundings (Art. 3, para. II).

The process of re-allocating farm production factors includes both the actual work and also the buildings, and often involves aspects which are intangible and qualitative. To this regard, it is interesting to recall that in 2000, the Basilicata Region of Italy passed a law establishing an environmental quality label for eco-compatible tourism and accommodation facilities, covering both business practices, as well as the actual buildings and the choice of building materials (L.R. No. 33/2000).

The same is true for Agriturist, the Italian Agritourism Association promoted by the Confagricoltura (General Confederation of Italian Agriculture), which launched a voluntary quality certification system in 2003. The Association's research has found that guests prefer a very definite type of accommodation: carefully restored and comfortably furnished buildings, a wide choice of farm products on offer for *in-house* catering, cultural activities connected with valorisation of agriculture and the environment, warm family-type hospitality (retrieved October 18, 2009 from www.agriturist.it).

Finally, referring to Schumpeter's innovative sources of supply, it may be seen that the agritourism farms are involved in a form of *economical farming* and tend to use internal production factors, limiting their use of outside inputs to the bare minimum and so reducing their costs (Henke and Salvioni, 2008). The innovative significance of this practice is defined in relation to the sectoral, productivist development model, where economies of scale predominate rather than economies of scope, and this means intensive monocultures, vertical integration, farmers deprived of their traditional expertise and skills in order to conform to the standardized globalized values imposed by *agribusiness*.

In this context the behaviour of the entrepreneur can no longer be that of the maximizing neo-classical entrepreneur, but is more adaptable because he realises that he is managing - by choosing diversification - a situation which falls within a model of "second best" (Lipsey and Lancaster, 1956). When farmers develop environmental concern, they tend not to maximize the net product per unit of surface area and the farm entrepreneur's objective function acquires parameters connected with the breadth of the multifunctional activity created.

9 Agritourism and multifunctionality

When Italian law interprets and regulates agritourism, it considers the implications of the multiplicity of the rural situation and the breadth of the concept of multifunctionality. This is in line with the most expert writings on agricultural economics and rural sociology and also with the orientation of EU (European Union) policy.

However, it is fundamentally important to clarify one point. The Italian legal system makes a very precise distinction of agritourism from all the forms of rural tourism, which come under the regulations for tourist and commercial activities. Agritourism is considered as an agricultural activity and its legal and economic *status* is based on its position in the sector of agricultural production. Agritourism offers typical products, services and hospitality, cultural and leisure opportunities, all managed by the farmer himself, in connection with the more traditional activities of cultivating the land, forestry and animal husbandry (Art. 2, para. I, Law No. 96/2006) without changing the agricultural purpose of the land and buildings used for agritourism.

The agricultural nature of agritourism is further clarified by the stipulation that the farmer himself is the subject operating the agritourism business (Art. 2, para. II, Law No. 96/2006). Therefore "agritourism is not just any kind of tourist initiative in a rural context, but is only that which is managed, singly or in association, by a person who is already a farmer" (Albisinni, 1983). At this point it is useful to take a systematic approach and refer to Article 2135 of the Civil Law Code in order to discover the qualifying characteristics of the farmer.

There is quite an interesting surprise here, because Article 2135 is very wide-ranging and has been up-dated recently by a Parliamentary Decree (D. Lgs. No. 228, May 18, 2001) concerning the orientation and modernization of the agricultural sector. This is a reformulation which dispels any doubt as to the essence of "*agriculturalness*" in agritourism, because it places agritourism unequivocally among the activities connected with the main activities of cultivation, forestry and animal husbandry which are typical of farmers (Art. 1, para. I, D. Lgs. No. 228/2001).

The connected activities are only those activities linked to one of the main agricultural activities: the connection has to be referred to the subjective extent in that it refers to activities which must be managed by the farmer; and to the objective extent in that they must involve the use of equipment and resources belonging to the farm (Minozzi and Pianelli, 2009).

This connection increases and becomes very innovative when the activity of agritourism mostly uses products of one the main agricultural activities, and when services are provided mostly by using the buildings and resources of the farm (Art. 1, para. I, D. Lgs. No. 228/2001).

The process of intangible innovation elaborated by the lawmaker expresses the criteria of prevalence clarified above. This underlies the capability of agriculture to operate as a modern production sector, able to open up, in its internal structure, processes of product diversification and operational multifunctionality, and to follow or at least to approach the model of rural, systemic, diversified and locally sustainable development to which the present work has referred several times, and which is the model towards which the present EU policies are directed.

10 Conclusions

The theoretical and developmental process described suggests a series of reflections. The first reflection leads us to look at the divergence between the theoretical analysis of the developmental problems of rural areas in the EU and the orientations of policy implemented. There is an interesting theoretical debate between academics engaged in further analysis of these themes, and the agricultural policy analysts. Not only this, but the EU itself has been the source of fundamental analysis and consideration, contained in documents which demonstrate a very advanced level of political and cultural sensitivity. For example there is Andriessen's Green Paper (1985), "The Future of the Rural Society" (EEC, 1988), the Cork Declaration of 1996, and the "European Development Framework" (EU, 1999), a very interesting document, though informal.

Policy definition does not show the same level of sensitivity about problems nor is there the same degree of analytical thoroughness about the ability to differentiate and to gauge aims and means in relation to the specific problems. In this regard, the course laid out by Agenda 2000 represents a meaningful turning point in the CAP, but this is still a little behind the Cork Declaration. In particular, rural development is the second pillar of the Common Agricultural Policy, but has not been given the financial support which the Commission itself announced, and the planning stage for 2000-2006 has remained rather tied to the logic of sectors and businesses, which never really died out.

The Mid-Term Review of 2003 has consolidated the process of reforming rural development, and, at the same time, deepened the political trend previously set by Agenda 2000. That is to say that rural development policy has been disconnected from any possibility of interaction with the policies of regional cohesion and brought back within the general CAP framework (Mantino, 2008). This choice creates widespread alarm about the allocation of financial resources to the first three axes of the 2007-2013 rural development policy. The fear is that at the following stage of regional negotiation, beyond the minimum quotas imposed by the EU, pressure from professional associations could lead to use of the resources for the second axis (environmental and countryside improvement), and the resources for the third axis (economic diversification and the quality of life in rural areas) being diverted to support the first axis (improving competitiveness of the agricultural and forestry sectors) (Musotti, 2007). This would mean a slide back into the old logic of sectoral productivity, and away from the criteria of integrated systemic development for the regional areas.

This leads into a second consideration which is that the model of diversified and multifunctional agriculture concerns all the types of farms in the European rural world, whether they are family-run or not. This formula does not, then, identify only the situation of marginal non-professional farms, without dynamic market prospects. It regards an alternative way of reconciling economic needs with ecology, social sustainability, the conservation of cultural and oeno-gastronomical heritage. This is an innovative model of agricultural development, supported by both EU and national agricultural policies and does not represent the antithesis of intensive agriculture, but rather integrates with and complements it (Henke and Salvioni, 2008). "What seems to be definitively outmoded is not, therefore, productivism itself, but rather the old model of mono-functionality" (ibid.).

At this point, the future of research is to carry out suitable methodological and quali-quantitative inquiries in order to evaluate the degree of productive diversification and multifunctionality which the European farmers are able to develop in order to meet the demand for commodity and non-commodity outputs which society requires of the primary sector. A flexible capacity for diversification and for "re-conceptualising rural resources as countryside capital" (Garrod et al., 2006), as in the case of rural tourism, will decide the direction that development will take. This is not unimportant, because this direction will in turn determine how fairly we treat and safeguard future generations.

References

Albisinni, F. (1983). Regioni e Agriturismo. In *Ministero dell'Agricoltura e Foreste*, Agriturismo 1983, Roma.

Bianchi, R. (2001). *Linee di pianificazione e programmazione economica in agricoltura. Dal piano di bonifica alla programmazione negoziata.* Bari: Edizioni Quadrifoglio.

Bianchi, R. (2004). *Bed and Breakfast: ipotesi di un'innovazione di prodotto tra categorie teorico-concettuali e prassi politico-istituzionali.* Bari: Edizioni Quadrifoglio.

Bianchi, R. (in press). Sviluppo rurale e prospettive locali: proposta di una lettura teorico-evolutiva, *Economia e Ambiente.*

Botta, F. (2001). *Della globalizzazione dello sviluppo sostenibile e degli altri giochi possibili.* Bari: Cacucci.

Brunori, G. (2003). Sistemi agricoli territoriali e competitività. In Casati, D (Ed.), *La competitività dei sistemi agricoli italiani.* Proceeding of XXXVI SIDEA conference. Milano: Franco Angeli.

CEE (1985). Prospettive per la politica agraria comune. In EU (Ed.) *Il Libro Verde della Commissione.*

CEE (1988). *Il futuro del mondo rurale.* European Commission, Com/88/501 def.

Consiglio Informale dei Ministri responsabili della gestione del territorio (1999). *Schema di sviluppo dello Spazio Europeo. Verso uno sviluppo territoriale equilibrato e durevole del territorio dell'Unione Europea.* Potsdam, May 1999.

De Benedictis, M. (1981). Contraddizioni e conflitti di interesse nella politica agricola comune, *La Questione Agraria*, 1.

De Filippis, F. (2008). Oltre il 2013. Il futuro delle politiche dell'Unione Europea per l'agricoltura e le aree rurali, *Quaderni del Gruppo 2013.* Roma: Tellus.

Di Iacovo, F. (2003). (Ed.) *Lo sviluppo sociale nelle aree rurali.* Milano: Franco Angeli.

Fabiani, G. (1986). *L'agricoltura italiana tra sviluppo e crisi (1945-1985).* Bologna: Il Mulino.

Fabiani, G. (1991). (Ed.) *Letture territoriali dello sviluppo agricolo.* Milano: Franco Angeli.

Fanfani, R. (1996). *Lo sviluppo della politica agricola comunitaria.* Roma: Carocci.

Garrod, B., Wornell, R., and Youell, R. (2006). Re-conceptualising rural resources as countryside capital: The case of rural tourism. *Journal of Rural Studies 22*(1), 117-128.

Graziani, A. (1998). *Lo sviluppo dell'economia italiana.* Torino: Bollati Boringhieri.

Henke, R., and Salvioni, C. (2008). Multifunzionalità in agricoltura: sviluppi teorici ed evidenze empiriche, *Rivista di Politica Agraria, Anno LXIII, 1*(Marzo). Retrieved October 18, 2009, from http:// www.agriturist.it.

Josling, T. (1993). Dopo l'Uruguay round: l'agricoltura nel nuovo ordine mondiale, *Rivista di Politica Agraria, 1*(1993).

Lipsey, R. G., and Lancaster, K. (1956). The General Theory of Second Best. *The Review of Economic Studies*, *24*(1), 11-32.

Mantino, F. (2008). *Lo sviluppo rurale in Europa. Politiche, istituzioni e attori locali dagli anni '70 ad oggi.* Milano: Edagricole.

Mantino, F., and Pesce, A. (1996). L'applicazione della riforma dei Fondi strutturali nelle aree rurali italiane, *Rivista di Politica Agraria*, Anno XIV(6), 3-15.

Meyer, H. von (1983). Wirkungslose Umweltpolitik- umweltwirksame Agrarpolitik. Überlegungen zum Verhältnis beider Politiken in der Europäischen Gemeinschaft. *Zeitschrift fur Umweltpolitik*, *6*(4), 363-387.

Minozzi, R., and Pianelli, L. (2009). *L'impresa agrituristica.* Santarcangelo di Romagna: Maggioli Editore.

Musotti, F. (2007). La politica europea di sviluppo rurale alla luce (anche) di alcuni insegnamenti di Federico Caffè. *Agriregionieuropa*, 3(10), 1-4 Retrieved August 23, 2009, from http:// www.agriregionieuropa .univpm.it/dettart.php?id_articolo=249.

OECD (1987). *National policies and agricultural trade.* Paris: author.

Ploeg, J. D. Van Der (2006). Esiste un nuovo paradigma di sviluppo rurale? In Cavazzani, A., Gaudio, G., Sivini, S. (Eds.), *Politiche, governance e innovazione per le aree rurali.* Napoli: INEA, ESI.
Ploeg, J. D. Van Der (2006). *Oltre la modernizzazione. Processi di sviluppo rurale in Europa.* Soveria Mannelli (CS): Rubettino.
Ploeg, J.D. Van Der, et al. (2000). Rural development: from practices and policies towards theory *Sociologia Ruralis*, *40*(4), 391-408.
Priebe, H. (1985). *Die Subventionierte Unvernunft*. Berlin: Siedler.
Rullani, E. (1997). Più locale e più globale: verso una economia postfordista del territorio. In Bramanti, A., and Maggioni, M. A. (Eds.), *La dinamica dei sistemi produttivi territoriali: teorie, tecniche, politiche.* Milano: Franco Angeli.
Saccomandi, V. (1988). Situazione agricola internazionale e riforma della PAC, *La Questione Agraria*, 29.
Schumpeter, J. (1971). *La teoria dello sviluppo economico*, (Introduction of Sylos Labini, P.). Firenze: Sansoni.
Schumpeter, J. (2002). *La teoria dello sviluppo economico*, (Introduction of Talamona, M.). Milano: ETAS.

Legal references

D. Lgs No 228 of May 18, 2001, n. 228 Orientamento e modernizzazione del settore agricolo, a norma dell'art. 7 della legge 5 marzo 2001, n. 57, Gazzetta Ufficiale 15.06.2001, n. 137.
EC Reg, No 1782/2003 of 29 September 2003 establishing common rules for direct support schemes under the common agricultural policy and establishing certain support schemes for farmers and amending Regulations (EEC) No. 2019/93, (EC) No.1452/2001, (EC) No. 1453/2001, (EC) No. 1454/2001, (EC) 1868/94, (EC) No. 1251/1999, (EC) No. 1254/1999, (EC) No. 1673/2000, (EEC) No. 2358/71 and (EC) No. 2529/2001, Official Journal of the European Union L 270, 21/10/2003.
EC Reg. No. 1698/2005 of 20 September 2005 on support for rural development by the European Agricultural Fund for Rural Development (EAFRD), Official Journal of the European Union L 277, 21/10/2005.
EEC Reg. No. 214/84 of 18 January 1984 amending Regulation (EEC) No. 2615/80 instituting a specific Community measure contributing to the development of certain French and Italian regions in the context of Community enlargement, Official Journal of the European Communities L 027, 31/01/1984
EEC Reg, No. 797/85 of 12 March 1985 on improving the efficiency of agricultural structures, Official Journal of the European Communities L 093, 30/03/1985.
EEC Reg. No. 2088/85 of 23 July 1985 concerning the integrated Mediterranean programmes, Official Journal of the European Communities L 197, 27/07/1985.
EEC (1987) Single European Act 1986, Official Journal of the European Communities No. L 169/1, 29.06.1987.
EEC Reg. No. 1760/87 of 15 June 1987 amending Regulations (EEC) No.797/85, (EEC) No. 270/79, (EEC) No. 1360/78 and (EEC) No. 355/77 as regards agricultural structures, the adjustment of agriculture to the new market situation and the preservation of the countryside, Official Journal of the European Communities L 167, 26/06/1987.
EEC Reg. No. 1094/88 of 25 April 1988 amending Regulations (EEC) No. 797/85 and (EEC) No. 1760/87 as regards the set-aside of arable land and the extensification and conversion of production, Official Journal of the European Communities L 106, 27/04/1988.
EEC Reg, No. 2052/88 of 24 June 1988 on the tasks of the Structural Funds and their effectiveness and on coordination of their activities between themselves and with the operations of the European Investment Bank and the other existing financial instruments, Official Journal of the European Communities L 185, 15/07/1988.

EEC Reg. No. 2078/92 of 30 June 1992 on agricultural production methods compatible with the requirements of the protection of the environment and the maintenance of the countryside, Official Journal of the European Communities L 215, 30/07/1992.
EEC Reg. No. 2080/92 of 30 June 1992 instituting a Community aid scheme for forestry measures in agriculture, Official Journal of theEuropean Communities L 215, 30/07/1992.
Italian law December 5, 1985, n. 730, Disciplina dell'agriturismo, Gazzetta Ufficiale 16.12.1985, n. 295.
Italian law Februar 20, 2006, n. 96, Disciplina dell'agriturismo, Gazzetta Ufficiale 16.03.2006, n. 63.
Italian regional law April 5, 2000, n. 33, Istituzione di un marchio di qualità ambientale per le strutture ricettive turistiche di Basilicata, Bollettino Ufficiale della Regione Basilicata, 10 aprile 2000, n. 25.

Part II

Food, wine and tourism

Educational farms in the Emilia-Romagna region: their role in food habit education

Maurizio Canavari[a]
Carrie Huffaker[a]
Rossana Mari[b]
Domenico Regazzi[a]
Roberta Spadoni[a]

[a] Alma Mater Studiorum – University of Bologna
Department of Agricultural Economics and Engeneering (DEIAgra)
{maurizio.canavari, domenico.regazzi, roberta.spadoni}@unibo.it

[b] Italian Region of Emilia-Romagna – Production valorization
rmari@regione.emilia-romagna.it

Abstract

Educational farms are farms which offer educational tours to visitors so that they can learn about specific information concerning agriculture, the environment and nutrition. These enterprises are involved in activities which complement their main work. This can bring benefits, including economic ones, to them as entrepreneurs, as well as having important social value. The objective of this paper is to present an authentic local example of this: the Italian Emilia-Romagna region, where Educational farms are particularly widespread and supported by the public administration. In particular, it highlights the role played by the educational farms regarding food habits and discusses the relationship between the enterprise's educational function and business strategies. This analysis is an attempt to view the above from a marketing point of view, highlighting its features and reporting on potential factors regarding improvement and development strategies. After a brief description of the general situation, we present data available from regional sources as well as from surveys conducted by private entities and by DEIAgra. The educational farms' role in regional education policies regarding food consumption will be highlighted. Finally, we offer some considerations concerning possible analysis in accordance with the framework of the marketing concept.

Zusammenfassung

Schulbauernhöfe sind Bauernhöfe, die Führungen für Besucher anbieten, um ihnen bestimmte Informationen über Landwirtschaft, Umwelt, und Ernährung näherzubringen. Diese Betriebe bieten ihren Besuchern Aktivitäten an, die die Kernaufgaben des Betriebs ergänzen. Dies bietet den Betrieben zum einen ökonomische Vorteile, zum anderen wird ein erheblicher sozialer Wert geschaffen. Ziel dieses Beitrags ist es, ein authentisches lokales Beispiel solcher Aktivitäten darzustellen: die italienische Emilia-Romagna Region, wo Schulbauernhöfe besonders weit verbreitet sind und von der öffentlichen Verwaltung unterstützt werden. Insbesondere wird die Rolle der Schulbauernhöfe bezüglich der Essgewohnheiten

K.L. Sidali et al. (eds.), *Food, Agri-Culture and Tourism: Linking local gastronomy and rural tourism: interdisciplinary perspectives*, DOI 10.1007/978-3-642-11361-1_5,

hervorgehoben und die Beziehung zwischen der pädagogischen Funktion und der Geschäftsstrategie des Unternehmens diskutiert. Diese Analyse versucht, die oben aufgeführte Situation aus einer Marketing-Perspektive zu betrachten. Sie zeigt ihre Charakteristika auf und berichtet über mögliche Faktoren für Verbesserungs- und Entwicklungsstrategien, die beachtet werden sollten. Nach einer kurzen Beschreibung der generellen Situation werden Daten aus regionalen Quellen präsentiert, sowie von Befragungen durch privaten Institutionen und durch DEIAgra. Die Rolle der Schulbauernhöfe in der regionalen Erziehungspolitik bezüglich des Lebensmittelkonsums steht in dieser Arbeit im Fokus. Zum Schluss werden einige Überlegungen bezüglich möglicher Marketingstrategien formuliert.

Riassunto

Le fattorie didattiche sono aziende agricole che offrono agli ospiti percorsi didattici allo scopo di far conoscere aspetti specifici legati all'impresa agricola, all'ambiente e all'alimentazione. Tali imprese svolgono attività complementari a quella principale, che possono apportare vantaggi, anche di tipo economico, all'imprenditore, ma che assumono una valenza sociale importante. Obiettivo di questo lavoro è quello di presentare una realtà territoriale italiana, cioè la regione Emilia-Romagna, dove le fattorie didattiche sono particolarmente diffuse e supportate dall'ente pubblico. In particolare, evidenzieremo il ruolo svolto da tali fattorie nell'educazione alimentare e discuteremo del rapporto esistente tra la funzione educativa e le strategie imprenditoriali dell'azienda. L'analisi vuole rappresentare un tentativo di leggere questa realtà in un'ottica di marketing, evidenziandone le caratteristiche e mostrando quali possono essere i fattori da considerare nelle possibili strategie di miglioramento e sviluppo. Dopo una breve descrizione del contesto di riferimento, si presenteranno i dati disponibili, provenienti da fonti regionali, da indagini svolte da enti privati e dal DEIAgra ed elaborati direttamente. Verrà evidenziato il ruolo delle fattorie didattiche nelle politiche regionali di educazione ai consumi alimentari. In conclusione si riporteranno alcune considerazioni relative ad una possibile analisi secondo le componenti del marketing concept.

Keywords: Educational farms; Multi-functionality; Nutrition education; Marketing concept.

1 Introduction

Educational farms are farms which receive visitors for educational purposes, where they learn one or more specific aspects of its business operations, about the rural and natural environment around the farm and about issues regarding food, culture and the rural tradition. This is one of the possible choices that farmers have when moving away from the single idea of agricultural production (the economic perspective) and adopting an innovative way to "cope with the negative aspects of productivity, or the agro-industrial paradigm [the sociological perspective]" (Henke and Salvioni, 2008). The farm enterprise adopts multi-functionality (OECD, 2001), and offers, in addition to supplying products, a number of other benefits and environmental services such as management of the landscape, the land itself, biodiversity, as well as social benefits such as the maintenance of rural societies, traditions, the safety and wholesomeness of food (Gaudio and Trunzo, 2009). Multi-functionality can be interpreted in various ways: a strategy for increasing the autonomy of businesses and the profitability of their resources, a way to combine income from agricultural production with that from environmental and social functions and a system leading to the involvement of the environment of which the farm business is part, as well as promoting sustainable development (Marsden and Sonnino, 2008).

In Italy, the multi-functional role of agriculture has been recognized in legislation with the enactment of Legislative Decree No. 228 of May 18, 2001 in implementation of the so-called "Orientation Law". It provides a new legal and functional form to agricultural enterprises, expanding the range of activities which can be defined as agricultural (Salghetti, Ferri, and Manghi, 2007). The farm enterprise, while remaining anchored to the agricultural sector, may perform activities which encroach on the industrial and service sectors, delivering services to individuals and society at large, and the occupation of farming is transformed by combining the functions of entrepreneur, farmer and rural technician and animator. The activities included in the concept of multi-functionality make the farm take on a different role regarding the market, and the strategies already, or still to be, adopted may be successfully reported and analyzed with a view to marketing.

To proceed with this approach it is necessary to try to define which activities may be attributed to multi-functionality in farming. Van der Ploeg and Roep (2003) make one of the first attempts to classify the activities which fall within the concept of multi-functionality. They report three categories of activity: deepening (broaden and enhancement of agricultural production), broadening (enlargement of the productive activities undertaken by farmers) and re-grounding (reallocation of production factors outside the farm) (Henke and Salvioni, 2008). The activities which take place on educational farms may be categorized as 'broadening'. In this case, in addition to its direct source of income, farmers see in educational farms an opportunity to promote the enterprise and its products (e.g., direct selling) or to provide motivation to young people and women not to abandon the countryside (Nasolini, 2005). In fact it stimulates the creation of new skills, enabling the development of cultural and educational attitudes in the group running the enterprise. The main activity is education, which is distinct from other activities falling under the name of multi-functionality because of its direct influence on human resources and its 'jointness' regarding not only technical aspects but mainly tied to the institutional behaviour of the farmer (Ohe, 2009). Educational farms, especially in their relationship with schools, have become an important tool for projects involving the environment, bringing together a system of values and educational disciplines (knowledge and skills) (Bertacci, 2005). There are various subjects which may become educational activities and often their choice depends not only on business and regional characteristics, but also on the guidelines dictated by public bodies operating in area. Emilia-Romagna has strongly promoted education regarding food consumption and knowledge about food (provenance, ownership and profiles). This approach is consistent with what is happening worldwide, and assumes major significance in this age where lack of information and bad eating and physical activity habits, are resulting in high costs to society. Diseases such as obesity cause 1 million deaths and 12 million cases of illness in Europe, generating high public costs; indeed, in some European countries the cost for treating obesity has reached as much as 1% of gross domestic product and accounts for 6% of direct health expenditure. In Italy, the phenomenon is growing ever faster and an average of 4% of children and adolescents are obese, while over 20% are overweight (Branca et al., 2007). In industrialized countries, governments have launched campaigns to raise public awareness, promoting healthy eating habits and tools for educating especially the younger generation in food awareness. It is precisely the younger generations which are most affected by their distance from agricultural reality, resulting in important gaps in their knowledge which affect their choices regarding food. The European Council of Young Farmers (CEJA) carried out a survey among children in Europe in the age group between six and eleven which revealed that children have mistaken factual beliefs about the food they consume (retrieved October 10, 2009, from www.ceja.org). They are influenced by advertising and their lack of food-related experience into having difficulty in associating original products with their final form after processing. For example, 50% of European children do not know where sugar comes from; three quarters do not know where cotton comes from, a quarter believing that it grows on sheep. A third of children are not able to quote even a single product derived from sunflowers. Most children come into contact with agricultural products only in the supermarket, with only 10% citing the farm as a regular source of supply for the family (of course these percentages vary in different countries according to the particular characteristics of their environment). Nearly a quarter of children are not able to cite a method of food preservation other than freezing. When asked which animals produce milk, all children mentioned cows. Strong differences appear, however, when children are asked what products can be obtained from milk. Cheese is the most cited, by three out of four children. Over half the children mentioned butter and yogurt and four out of ten cream. Other products (whipped cream, cheese curds, ice cream ....) are cited less frequently (Holst, 1999).

Against this background, we have tried in this paper to highlight the relationship on educational farms between food education (including correct food consumption) and the development strategies of the enterprise itself. The study proposes firstly to focus attention on how, in fact, educational activities integrate with those of the 'traditional' enterprise, using an approach based on the marketing concept. We believe that this approach is useful for analyzing the relationship between their prospects for business diversification and their educational activities, up to now little explored in the literature (Ohe, 2009). Emilia-Romagna, which happens to be one of the Italian regions most active in promoting these issues, was taken as a reference point. Below we briefly describe the characteristics of educational farms and their evolution, and some data will be given on current national and regional situations (focus on Emilia-Romagna). It will highlight the role played by the educational farms in consumer education projects developed in the Emilia-Romagna region, concluding with some considerations regarding the system of educational farms with regard to the marketing concept.

2 Educational farms

Educational farms have existed in Italy for less than 15 years but have, however, been around a lot longer in other countries. The precursors in this field were the Scandinavian countries: Norway, Denmark and Sweden put into practice in the early 1900s the ideas of an American movement, the 4H Club (Head, Health, Heart, Hand), which described its aims as the harmonious development of the individual and the concept of 'learning by doing'. Thus 'animation' or 'city farms' were established: Farms located in urban or suburban areas and established with the aim of bringing together children, teenagers and adults in the city with farm animals, soil and wild and cultivated plants. They were usually set up on land and in buildings owned or operated by community leaders and volunteers from non-profit associations. From these three countries it then spread to other nations: in Germany after World War II, in the Netherlands, where they are also sustained by public and private bodies, in the United Kingdom where 'city farms' and 'community gardens' carry on a wide range of activities such as training courses, visits by school groups, horse-riding and vegetable-garden therapies, composting and recycling of waste, leisure activities and sports, summer camps, after-school clubs, etc. Similar activities have also been recorded in Flemish and French-speaking Belgium. In France, they are specially supported by a public state body. In Spain, the legislation which regulates the activity does not allow the farmer to be directly involved in teaching, requiring the support of community leaders and experts who are graduates in educational sciences (ALIMOS, 2009).

The first Italian Educational farms were set up in the late 1990s, and have grown steadily since 2000. In the same period, besides the Cesena Agri/environmental Observatory network also the Mantova Agritourism Consortium and the Piedmont Agri-tourism Consortium kicked off, soon followed by Emilia-Romagna, which in 1999 promoted networks of educational farms in all its provinces.

The regional 'educational farms project' evolved from the 'open farms project'[1], the only one of its kind in Italy, both regarding the territory it covered (the entire region) and regarding the synergy created between public and private institutions (Mari, 2009). The main purpose was to promote among children and young people a culture of food, understood as the connection between nutrition issues and their relationship with the land and local tradition. Educational farms are in fact defined as "agricultural enterprises and associations which, in addition to carrying out traditional agricultural activities, including educational activities targeted at various age-groups of schoolchildren and other types of users, are aimed at:

(a) developing knowledge of the countryside, agriculture and its products and in general the link between diet and cultural and historical heritage;
(b) educating consumers through an understanding of the relationship between production, food consumption and the environment with sustainable development as an end view;
(c) spreading knowledge of the biological cycles of plants and animals and the production, processing and preservation processes of local agricultural

[1] As for "open farms", this initiative has been organized since 1999 and in recent years annually more than 200 agricultural enterprises have joined and over 60,000 visits registered in the opening two days (ten times the attendance recorded in the first year of the initiative). Over 50% of participating enterprises are also educational farms. Here, programs to discover the cultural sights of the area, trekking and cycling have been organized and offered to the public, as well as visits to the farms.

products in relation to agricultural activities conducted on the farm." (Regional law No. 4 of March 31, 2009).

Educational farms can offer services to a wide variety of target groups (schools of all grades, groups with special needs, families, individuals, etc.) through, for example, therapeutic activities aimed at disabled and elderly people (e.g., pet-therapy), offering cultural and scientific programs to attract users from higher educational levels and offering courses of all kinds (e.g., cooking, processing and decoration of clay, etc.). Besides offering these activities there might be the direct sale of farm produce, which is developing in new ways: opening a store in the country or the nearest city to allow customers greater convenience when buying, taking part in fairs and local markets to sell and promote both the products and the enterprise, selling via Internet, and promoting initiatives like 'pick your own', a form of direct marketing in which consumers/customers pick fruit and vegetables themselves. Finally, educational farms can join in with other initiatives such as oeno-gastronomical routes, i.e., itineraries with high tourist potential, characterized by high quality local products which are placed in a setting of scenic, historical and artistic activities and which enhance the image of the company.

A new regional law (No. 4/2009) lays down rules on training (consistent with types of production and closely linked to the orientation of food consumption), procedures for enrolment in the provincial list of approved enterprises and enterprise set up declarations, structural requirements, sanctions, monitoring activities and training. A pivotal element of the project is the training of the enterprise manager which confirms the obligation, both at the time of becoming a member and in successive years to continue training (Mari, 2007). In this training process, great emphasis is given to innovative methods of education, in particular food education. It should be noted that while in the early days of the project activities were initiated as a support for schools, there has been a great increase in the tendency for other users to come to the educational farm to get to know, buy and try traditional products, local specialties as well as to experience the local rural culture, and so on.

2.1 Some national and regional data

Data on educational farms are still fragmentary at the provincial level and the only attempt to organize them was the census carried out by ALIMOS in 2009. Based on this source and integrating it with investigations carried out directly, it appears that there are 1,878 authorized educational farms throughout Italy. They were 258 in 2000 and less than one third of the current number in 2005. The regions with the highest concentration of educational farms are Emilia-Romagna and Campania, with 330 and 245 enterprises respectively. The importance of these enterprises in the total national agricultural and regional picture is extremely limited (only one per thousand, about the same in all the different regions), but if one compares the data regarding agritourism enterprises it is clear that in some cases individual regions differ. Regions strongly linked to agritourism, such as Trentino, Tuscany and Umbria are not the same with regard to educational farms. Furthermore, it is notable that on average, the ratio between teaching and tourist farms is about one to ten, while in some regions it is about one to three, as it is for example in Emilia-Romagna, Campania and Piedmont. Of course, one should remember that often the educational farms are also tourist farms, even without this connection being required.

Concerning the situation in the Emilia-Romagna region, information is derived from manifold sources:

- an analysis carried out directly on data reported in the 'Guide to Educational Farms in Emilia-Romagna' (RER, 2007);
- ALIMOS data of October, 2009;
- a study performed on behalf of the Emilia-Romagna Regional Government by the former Business Research Centre (now Dinamica) in collaboration with the former Cesena agro-food Centre (now ALIMOS). The survey addresses the quality plan of educational farms, trying to identify the needs of the farm operators running such enterprises in order to improve their services. It is based on telephone interviews with all the companies in Emilia-Romagna which adhere to the 'quality charter'[2] and are thus part of the regional circuit. Furthermore, a group of 247 companies distributed over nine provinces were interviewed directly (Cavazzuti and Macario, 2005);
- a study performed on behalf of the Emilia-Romagna Regional Government by the former Business Research Centre (now Dinamica) in collaboration with the former Agro-environment Observatory (now ALIMOS). The survey aimed at assessing the level of application of the requirements contained in the 'quality charter' and was conducted directly by the provincial food education authorities. The data were then processed at regional level, which was carried out by the Agro-environmental Observatory (years 2003-2004).

Despite not having all the data for past years, it is considered that the course charted by these sources is also valid for the current situation.

Since the beginning of the year 2000 the number of educational farms accredited by the regional government has increased steadily (except in the years 2007 and 2008) to reach the current number of 330. These enterprises are more concentrated in the provinces of Bologna and Modena (together representing over 38% of the regional total), followed by Reggio Emilia and Piacenza (each with about 11%) and their location is fairly evenly shared out between the plains and the hilly areas, while there are few in the mountain areas. The size of the enterprises varies widely, from 3 hectares up to 500 and above, with a large number being concentrated between 10 and 50 hectares (bigger than the 'regular' farms). Regarding the type of production, about 44% of the surveyed enterprises said that they used organic production methods and 27% dealt with products protected by Geographical Indication Labels (this was only 7% in early 2000).

---

[2] The "quality charter" is a tool which was thought of before the adoption of the more recent legislation on educational farms and had become a point of reference for formal approval and control by the public administration as well as a way to ensure quality services adequate to the aims of the project. It contained the basic requirements for submitting an application, the production characteristics of the company, the training requirements for farmers and finally the facilities and skills needed to ensure high education and safety standards.

Almost all enterprises have a classroom, a prerequisite for conducting business even in bad weather. Another important feature is the presence of marked-off safe areas for play and picnics; this is found in about 95% of cases. The number of people employed on the teaching side in the enterprises varies from 1 to 3; amongst those who work in the enterprise employed in teaching there is a spouse or partner in 30% of cases, another family co-worker in 27% , employees in 8%, employees with a professional contract 5% and professional teachers (for schools) 4%.

The person in charge of teaching appears to be predominantly a woman (over 50% of cases, with the highest percentage in the provinces of Bologna, Piacenza, Parma and Modena), while the entrepreneur is a man (62% including cooperatives and personal enterprises).These entrepreneurs have passed a basic training on educational farms and nearly half have attended an advanced training.

Regarding the initial motivation that prompted the contractor to begin this activity, ethical values (culture, environment, education about nutrition) were indicated by 86% of respondents, indicating a fairly universal agreement of farm operators to convey to the younger generations traditions, knowledge and tastes of the agricultural world. This was followed by increase of farm income (38%), keeping up with the times (21%), employment for family members (15%) whereas 29% of respondents gave other reasons, such as the promotion of already existing activities (e.g., rural tourism).

Almost all farm operators run other multi-functional services alongside their educational activities: 60% are involved in direct sales of agricultural products, although with considerable differences among the various provinces (this figure was about 75% in the early 2000s). The agricultural products mainly sold are fresh and processed fruits and vegetables, wine, balsamic vinegar, olive oil, cheese, meat and sausages, honey, propolis, royal jelly, liquors, cereals and medicinal plants. Furthermore, more than half participate in the "open farm" scheme and about 50% are tourist farms. Among the latter, the majority offers restaurant and overnight stay facilities. Finally, 30% are part of oeno-gastronomical itineraries.

Besides the increasing number of enterprises participating in the project there has also been an increase in the average number of groups per year visiting the farm, which in 2007/2008 was approximately 20 (ALIMOS, 2009). The total number of visitors has risen from less than 70,000 in the years 2002/2003 to the current 122,500.

For the same year (i.e., 2007/2008), the predominant type of user continued to be schools, ranging from nurseries to high schools (with a notable prevalence of primary schools), which over the last two years have hovered around 80%. This figure has changed significantly since the first survey (2003/2004) when it was 93%, since only 6% of visits were made by adults and other groups (ALIMOS, 2009). The visits take place mainly during the months of April and May (three quarters of the total) and the duration of the visit in over 50% of cases is half a day. The periods of being open to the public are not necessarily continuous throughout the year; sometimes they coincide with school terms, while in other cases they are open only in the spring and autumn months. The number of visitors accommodated per day varies greatly (from

10 to 80) and depends primarily on the number of people on the farm who devote themselves to teaching.

Regarding the educational programs which are run on the farm, almost all propose a guided tour. About 80% of the activities involve either the farm building or the farm animals (sometimes singly, sometimes in herds, while some do pet-therapy). Environmental protection is crucial and is offered in programs such as organic farming, composting and biodynamic agriculture. Also widespread is plant recognition which can be done for medicinal and aromatic herbs, fruits of ancient times, flowers, wild berries, etc. Play activities are offered, especially for younger children, as are theatre and tales of myths and legends. Also veritable museums of tools related to traditional crafts are set up. Activities which may be related to education about nutrition are provided in 85% of cases, such as 'From farm to table' or workshops regarding food. The courses are mainly about the production of bread, milk and cheese, meat, wine, oil, honey, canned fruit and vegetables. It should be noted that in the provinces of Bologna, Piacenza and Rimini these programs are less common (around 70% of the provincial total).

With regard to prices, however, there is some variability both between the various provinces and within the provinces themselves, except for some provinces (Parma, Reggio and Forlì-Cesena), where the price is standardized. The average cost ranges from a minimum of 4-5 Euros to a maximum of 8-9 Euros per visitor (the lowest and highest are 1 and 12 Euros respectively); the variability depends on the duration and type of course. In addition some companies, if not already included in the agreed cost, also offer snacks and lunches to buy (from 3 to 20 Euros).

The program must be prepared together with the teachers and the material provided can vary considerably; it may be produced by the farm itself or it may be prepared by the regional administration or by specific bodies. An evaluation sheet is provided for teachers to compile about the experience on the farm. This is an important and useful tool, primarily for farmers in order to continually improve the services they actually provide, and this form should always be provided as required by the 'Quality Charter'. This rule, however, is often not respected and the number of forms completed is small compared to the number of visits actually made.

As an example, we list below some of the information obtained from the analysis of these forms in the province of Bologna (50 sheets completed by teachers during the school year 2004-2005 with thirteen farms involved).

The most requested educational programs were:

1. from wheat to bread
2. getting to know farm animals
3. milk production
4. discovering the surrounding environment
5. "the 5 senses" program

6. education about nutrition
7. traditions and memories of the past

Whatever the activity on the farm, all visits are covered by a specific educational project which will be further developed in class and is directly or indirectly related to nutrition. The assessments made by teachers on a satisfaction scale ranging from (1) ‘not good at all’ to (5) ‘very good’ highlight the high level of coordination with the farmer in defining the program and the satisfaction of expectations (scored ‘very good’ by 76% of teachers). Assessment of the available tools and of the farmer’s skills were particularly positive in most cases, as were assessments regarding the usefulness of the activities carried out, which provided insights to be further developed at school. Both the level of understanding of the proposed issues and the students’ participation scored positively as well. The overall scores given to the overall experience at the educational farm were listed as follows: 79% gave 5 points, 13% gave 4, 6% gave 3 and 2% gave 2. Most were considering repeating the experience.

Among suggestions for improving the educational activities the following were cited the most:

- the creation of more informational material
- the organization of more ‘hands on’ activities for the children
- diversification of the types of activities proposed and, in some cases, better organization
- more accurancy in the presentation of the parts of the farm involved in the visit
- lower prices

2.2 The link with consumer orientation and the role of women

Initiatives in both the public and private sectors designed to steer consumers toward healthier behaviour are many and varied. The value of proper eating habits accompanied by adequate physical activity in the prevention of many diseases is now widely recognized (Ministry of Health and Social Policy, 2009 Program). The cornerstone of the success of these initiatives is how information and all forms of consumer education are disseminated. The importance of these activities is evident when one considers that recent studies have shown that obesity, which is more common in the poorer strata of society, is due not only to economic difficulties, but also and especially to the lack of a cultural environment which places value on, and uses its energies for, transmitting certain beliefs (Wills et al., 2008). This educational role must be played also by the school, a place where kids can develop and learn.

The Emilia-Romagna regional administration has for some time been promoting education on nutrition, creating projects and initiatives in schools in collaboration with the provincial administrations. Education about nutrition in Emilia-Romagna has moved away over the years from the concept of prevention and health protection, concentrating more on the concept of welfare and the development of agriculture as the primary source of food, culminating in 2002 in a specific regional law which brings together production, nutrition, environment and health. Law No 29 of November 4, 2002[3] draws up a framework of objectives and priorities within which all parties involved are called upon to take action, taking into account the many and positive experiences gathered over these years and fed by the conviction that starting from nutrition itself we can promote a better quality of life and health. More specific objectives, activities to implement and methods of assessment are defined in the "Program for consumer orientation and education about nutrition" renewed for three-year periods. The projects in this field are about getting across the right information regarding the relationship between agriculture and the environment, between diet and health, promoting the dissemination of historical, cultural and anthropological information related to food production and the lands of its origin and also working for behavioural change as well as promoting a real change in the nutrition of young consumers and their families.

Article 2 makes explicit reference to the role that the food-producing farms might have "in promoting food education programs [...] designed to develop [...] educational and informational activities". We must reiterate that educational programs focussing on nutrition are not an option or a potential activity but a constraint that the regional administration places on the educational farms that want to stay in the network. This aspect is specific to Emilia-Romagna, which in fact involves Educational farms in numerous food education initiatives (see Table 1). These are multi-year projects, involving different protagonists and with clear objectives. In a retrospective analysis of projects already carried out, the importance of initiatives emerged which require close cooperation between different protagonists (region, school, province, training and coordination institutions). This cooperation, however, is conditioned by a lack of dialogue, a lack of a common vision and inadequate organization considering the complexity of the situation. A clear and shared methodology was judged very positively by all parties involved in the projects. Educational farm managers in particular have stressed that such initiatives, seen as being very positive, consolidate relationships with other participants and that it is important to be able to operate even outside their own area. In this way the educational role of the farms is enhanced. There are complaints, however, about the sporadic nature of the initiatives, with no continuity in planning and lack of involvement of students' families.

---

[3] Law No 29 of November 4, 2002: "Rules for guidance on food consumption and education and assessment of catering services".

Table 1 Educational farms: role in food education initiatives

Label of the initiative	Period	Objective	Eating occasion	Targets	Activity type in the E.F.	No. of E.F.	Programs/ occasions/visits	Comment
Eating together	2002-2005	Consumer education	Breakfast, snacks, fast-food	Nurseries, primary, middle and high schools as well as parents and teachers	Site of visits, involvement in the creation of recipes	22	85 programs	The farm visit was appreciated mainly because it was part of a defined program
Tasty snacks	2004-2005	Consumer education, reviving flavours and ways of tasting	Snacks between meals	Primary and middle school	Food supply and traditional recipes	45	934 occasions (snacks)	The importance of trying new foods (different from industrial products)
Tutti frutti 5 portions, 5 colours	2005-2007	Encouraging consumption and knowledge about fruit. Choosing food wisely	Snacks	Primary school	Guided visits. Interactive workshops and suppliers of raw materials (discovery and appreciation of traditional products).	-	9000 children involved	Production of educational material changing eating habits; information from a questionnaire
The kitchen garden at school – Let's plant good food	2006-2008	Environmental education tied in with food education	-	Nurseries, primary, middle and high schools	Training by E.F. managers	14	109 schools	The importance of "learning by doing", focus on the local area

The commitment of the regional administration regarding food education was confirmed by the merging of the Directorate General for Agriculture and the Directorate General for Health, which took place in 2008. In particular, while implementing the regional plan for preventive measures at the Health Directorate, a working group was set up on 'Control and prevention of obesity', which involved also staff from the Directorate for Agriculture, with the aim of coordinating efforts for promoting the adoption of healthy lifestyles by the population. This collaboration was reflected in the drafting of "Regional Guidelines for school meals" and the identifying of common goals between agriculture and health authorities for activities regarding education about nutrition. The investigations conducted showed the role played by women in places like educational farms, open farms and tourist farms. "Current experiences indicate that women, starting out from their age-old, traditional roles, show themselves able to transmit and adapt to today's world the values of an agricultural culture developed over millennia, and to project them into the future" (Ballarini, 2004).

The original idea, meaning the initial impetus to get involved in educational farms, comes from diverse origins: some farmers created a real project, others got involved for the simple pleasure of doing so. Nine out of ten farmers would not exchange their job in the country for a steady job in town. Being involved in an educational farm, for a woman, offers great job satisfaction, but it is not in itself a substantial source of income. To compensate for this, it does make it easier to conciliate leisure time and working time; enabling women to choose their own timetable, to keep up to date with things and to use the computer and networks.

There are outstanding examples of multi-functional farms managed by women and there are cases of significant generational changes involving the children of farmers. These are often the bearers of skills, study qualifications and experience, also gained in other areas, who are capable of "generating" particularly attractive projects, services and proposals for hosting visitors on the farm, opening the company to local projects and participating in comprehensive programs for education about nutrition and the environment and addressing eco-sustainability issues (Stinga, 2004). For women multi-functionality is something lively, creative, full of ideas, leading to the discovery of new, more advanced ways for women to gain their identity, but also to learn how agriculture works.

3 Educational farms, education about nutrition and marketing strategies: some conclusions

Studies in the field highlight some distinctive features regarding the structure and activities of educational farms which provide food for thought about guidelines and tools for the development of these enterprises. Starting from their educational function, which in Emilia-Romagna is particularly linked to food education, it is necessary to use a broader approach to analyze future prospects. Indeed, the different activities on offer, and which are part of the multi-functionality of the farm, can be analyzed and developed using the strategic approach to marketing. The basic premise is that a multi-functional enterprise can make major economies of scope by offering educational services, but this can not be improvised and requires, as well as a clear justification for the entrepreneur, some specific investment and modifications regarding corporate structure and organization. Analysis of these factors may therefore be based on the phases of strategic and operational marketing, in which it is necessary to define missions, goals, strategies and actions.

To define the mission of educational farms one must take into account the real motives which move the entrepreneur to join the project. These motives are influenced by the personal history of the entrepreneur and the farm, a predisposition to getting on with people, the situation in question, real adherence to the governing administration's aims and finding the balance between the desire to offer a service and how much the market and society in general will pay for it. In recent years there has been a shift in motivation, mainly concerning educational activities for schools, which is more purely economic and is more about diversification and promotion of the enterprise and its services. In some cases this change is likely to distort the 'philosophy' in question (education in agriculture, food and the environment), to replace it only with an increase in income. Besides, it is unthinkable that these activities are carried out only by highly motivated entrepreneurs who operate in a 'voluntary' spirit without being able to count on a real financial return for the enterprise. It therefore becomes important to bring together these two components and try to define clearly the areas where the company wants to act and the strategies it wants to employ. One notes, for example, the spreading phenomenon of 'social farms', in which users belong to disadvantaged groups (people with disabilities, ex-addicts, former prisoners and young people with learning or social handicaps, etc.) and where the company's mission is rooted in specifically ethical grounds[4] (Franco and Senni, 2001); this service is a source of income for the company.

The *objectives* of entrepreneurs are highly sensitive to the particular features of the territory and socio-economic system where the farm is situated. It is for this reason that provincial, regional and national bodies express such greatly varying objectives, and this also creates difficulties in trying to coordinate efforts and improve the efficiency and the effectiveness of actions taken both by individuals, and by public bodies. If the only goal for opening an educational farm is to increase the farm's income, experience tells us that this is achieved only by companies that can accommodate a large number of guests (Mortini, 2003). In fact, they take advantage of the educational activities to promote more profitable services to the public such as agritourism and on-farm and off-farm direct sales of their products. In today's socio-economic environment enterprises express a variety of aims, but it is important to define priorities in order to be able to identify and make use of the available resources.

---

[4] More details are available at the following link: www.Fattoriesociali.it. The regional administration of Emilia-Romagna has included 'agriculture in social cooperatives' within the category of 'agritourism' (Law 4/2009).

The surveys considered above confirm this need, indeed the motivations expressed by entrepreneurs range from 'ethical' to 'income growth', but the 'success' of the initiative depends fundamentally on clear expression of objectives, the quality and quantity of resources made available and the type of entrepreneurial approach.

This approach becomes evident with the definition of *strategies* and relative *actions* considering the four elements of the marketing mix:

- *Product/service*: The product strategy concerns the choice of programs to be activated, the chosen target, the timing of the offer and synergy with other company activities. The choices depend on the specific nature of the enterprise regarding such things as the type of cultivation, the structure and infrastructure available, the rationale and objectives of the entrepreneur and the human resources involved and their skills. In particular, the success of the initiative is closely linked to the ability of the people who coordinate the activities at each farm to find their own way to making the most of their "rural world" (Napoli, 2003). The importance of the specificity of the offer should not, however, make us forget that we need to find a reference system which lays down standards of expertise which should be common to all firms participating in the project. Tools like the aforementioned "Quality Charter" are the 'ticket to ride'. This should be allied to a tool for defining the services offered and to indicate what the quality factors are, taking into consideration the different viewpoints of the people involved. Obviously, the service offered is strongly influenced by external constraints such as the cost of transportation (which falls on the school and families in the event that there is not already a school service available) or legislative constraints on business (serving of food and drink only possible for tourist farms, enforcement of safety standards, etc.). The regional situation in Emilia-Romagna has shown a variety of services, in most cases linked to the particular nature of the locality but sometimes going beyond this as well as beyond any educational aims. In some cases there is a risk that the educational farms may become something 'other' than what they were set up to be, concentrating more on other activities, thus taking away resources from the specific work of the farm and lowering the quality of the service offered and generating confusion among its users.
- *Distribution:* Here transport becomes the main issue. The product or service is not transported, but there is the problem of transporting visitors to the farm (as there is for direct sales on the farm). There are various solutions, some more costly to the producer, others more costly for the user. For example, organizing busses to collect visitors on set days and at scheduled times is a possible solution. But it must be indicated who bears the cost and what kind of assistance is necessary to manage it. Concerning distribution, there are also issues related to activities carried out directly in schools, such as the transportation and conservation of products.
- With regard to *price*, variables such as the users themselves, the type of location, the human resources, the materials needed, etc., all determine choices on this issue. We have also seen how within the specific region there are

provinces which have induced farms to accept agreed and standardized prices. Each choice has its positive or negative side (e.g., greater transparency, less diversification and restrictive constraints on the service). It remains the case that price is the factor which notably influences visitors' choices, especially for schools.

- *Communication* is as always a key factor in the success or failure of these kinds of initiatives and the types and means of communication can vary greatly. The choice depends primarily on the persons involved in the information flow which is generated, e.g., between public body (regional and provincial) and enterprise, among public institutions, between public body and schools/other groups, between enterprise and schools/other groups and between company and individual clients. The regional farms in particular complain about a lack of support from the public administration in promoting their activities in schools. Each province has sought to develop a set of documents/guides which carry information about participating farms, but their distribution is often limited. The region, in collaboration with designated agencies, has published educational materials and some documents which came out of specific initiatives to support business activities. Even enterprises do not pay particular attention to communication. According to the mentioned surveys the primary mode of communication and promotion is word-of-mouth and this is also evident if we analyze data on the frequency of visits to individual farms: some are visited more often, above all by classes from the same school or school district. The majority of enterprises do not implement specific initiatives, mailing is used in a clear majority of cases, but the web is not among the means used for promotion and contact with customers. In terms of communication the limited request for grants is notable (e.g., only 30% asked for and received assistance under the Emilia-Romagna Regional Rural Development Plan 2000-2006, measure 3.P "Diversification of agricultural and related activities", Action 2) and can also be put down to inadequate publicity.

In conclusion, the experience of educational farms, being mainly prized for their work as providers of educational services, must be analyzed as a system which is more complex than it might have been originally. Recognition on the part of the 'public bodies' and of the market regarding this 'educational and social role' is not always up to the expectations of the enterprises themselves. This is why we believe that educational farms should try to move towards a marketing approach in managing and enhancing their activities. More particularly, in the following we propose a possible list of 'quality/strategies': it is a check-list of factors to be used to analyze and improve the educational farm system. The basis of this proposal is the list of quality indicators proposed by Bertacci (2005), which is the result of reflections on the design and implementation of educational programs on farms in Emilia-Romagna.

To this list, which specifically regards the relationship between schools and farms with an underlying 'social impact', further considerations have been added that take into account business factors and issues regarding the different types of users. From the business point of view, the strategic factors of educational farms are:

- Quality and meaningfulness of the proposal. Quality and meaningfulness must be assessed in the light of the objectives to be pursued. Customer satisfaction (schools, groups, individuals, but also public bodies) have become important tools for analysis.
- Shared project planning and continuity over time. For schools, integration with the Training Plan (POF) makes the project part of the schools' objectives, giving them strength and continuity. For other types of users, the project could be part of individual recovery/support schemes for people in difficulty, or be part of social and cultural programs operated by public or private associations. An appropriate evaluation tool should focus on how many objectives have been achieved. The planning process also involves defining the operational phases of projects which would follow a common reference framework (description of the entities, roles, timing, type of equipment, necessary forms, etc.) and also include the active involvement of end-users of the project (students, teachers, social and health workers, etc.).
- Diversification of the proposals.
- Interactive approach, workshops, leisure and sensorial methodologies. The methods used are important strategic features in the success of the programs. Here analysis entails an evaluation of teaching materials, sometimes developed by external experts, or the ways innovative tools are used in teaching.
- Adequacy of the structure itself. Safety is certainly the most relevant and important factor for organizations which host schools and groups. Farm structures are not always appropriate, and the necessary investments could affect business results. The company must still aim to make the farm ever more welcoming and safe.
- Reliability and competence of the contact persons in the company. It is possible to improve the level of professionalism amongst entrepreneurs through training (regarding both general methodology and the teaching of specific topics).
- Visibility of the service and availability of information.
- Adequate communication.
- Reference systems and networking. In order to improve and diversify farms, to adopt voluntary standards of quality and definition of levels of service are proposed. Any analysis should also consider the public body which currently needs to improve and redevelop controls and its quality charter (this can be a valuable source for analysis of customer satisfaction and an assessment tool for enterprises). Improvement of the service can also be achieved by stimulating forms of collaboration between farms. Cooperation among enterprises can certainly be encouraged by establishing clearer rules and specifications (e.g., protection of the educational farm's name, single address regarding Value Added Tax exemption, etc.) and enabling them to take advantage of the opportunities offered by national and regional administrations (grants, business expansion opportunities, other forms of cooperation).

- Special projects and synergy with other activities. Initiatives such as 'school-farm twinning' (e.g., food education) and sales of related products and services (speciality food and catering) can help to achieve some economies of scope.

References

ALIMOS (2009). Le Fattorie Didattiche in Italia. Retrieved October 31, 2009, from http://www.fattoriedidattiche.net/.

Ballarini, G. (2004). Il futuro del passato. Il Divulgatore 9/10, 8-15.

Bertacci, M. (2005). La scuola e le fattorie didattiche. In Bertolini, S. (Ed.), Nuovi educatori ambientali/2-Esperienze seminariali nel Master in Educazione ambientale. Quaderni INFEA 4, Regione Emilia-Romagna.

Branca, F., Nikogosian, H., and Lobstein T. (2007). La sfida dell'obesità nella Regione europea dell'OMS strategie di risposta. Quaderni di Sanità Pubblica 2007(146), 3-85.

Cavazzuti, S., and Macario, M. (2005). Fattoria, chi sei? Il Divulgatore 7/8, 20-30.

Holst, C. (1999). Report of a representative survey on "Children's knowledge on agriculture" CEJA Education. Mölln: INRA, Deutschland.

Franco, S., and Senni, S. (2001). Agricoltura sociale e sviluppo rurale. Alcune riflessioni introduttive. In Carrà, G. (Ed.) Atti del XXXVIII CONVEGNO SIDEA "Servizi in agricoltura.", September 27-29, 2001. Catania: University of Catania.

Gaudio, G., and Trunzo, P. (2009). L'agricoltura sociale nella politica di sviluppo rurale in Italia. Atti del Convegno di Studi Rurali, Ripensare il rurale: nuovi bisogni, innovazioni e opportunità per lo sviluppo sostenibile del territorio, June 25-27, 2009, Altomonte (CS): INEA.

Henke, R., and Salvioni, C. (2008). Multifunzionalità in agricoltura: sviluppi teorici ed evidenze empiriche. Rivista Economia Agraria 1, 5-34.

Mari, R. (2007). Un sistema che cresce. Il Divulgatore 11/12, 10-15.

Mari, R. (2009). Introduzione, Educazione alimentare in Emilia-Romagna – Le tappe di costruzione di un sistema regionale. Bologna: Regione Emilia-Romagna e Dinamica.

Marsden, T., and Sonnino, R. (2008). Rural Development and the Regional State: Denying Multifunctional Agriculture in the UK. Journal of Rural Studies 4, 422-431.

Mortini, E. (2003). Le fattorie didattiche mantovane: aspetti economici e strategie di marketing. Retrieved October 10, 2009, from http://www.tesaf.unipd.it/pettenella/ tesi/Mortini.htm.

Napoli, L. (2003). A new reality for Italian rural areas: educational farms. Proceedings of the Conference: The cultural turn in geography, September 18-20, 2003 (pp. 333-345) Gorizia: Edizioni Universitarie di Trieste, pp.333-345.

Nasolini, T. (2005). Educare all'ambiente e all'alimentazione. In Bertolini, S. (Ed.), Quaderni INFEA (Vol. 4): Nuovi educatori ambientali/2-Esperienze seminariali nel Master in Educazione ambientale. Bologna: Regione Emilia-Romagna.

OECD - Organisation for Economic Co-Operation and Development (2001). Multifunctionality: towards an analytical framework. Paris: author.

Ohe, Y. (2009). Educational Function of Agriculture and Farm Diversification: Evidence from Dairy Farming Experience Services in Japan. Contributed Paper prepared for presentation at the International Association of Agricultural Economists Conference, August 16-22, 2009, Beijing, China. Retrieved October 31, 2009, from http://www.ageconsearch.umn.edu/ bitstream/51557/2/IAAE2009paper_Ohe.pdf.

Ploeg, J. D. Van der, and Roep, D. (2003). Multifunctionality and rural development: the actual situation in Europe. In Huylenbroeck, G. Van, and Durand, G. (Eds.), Multifunctional agriculture: a new paradigm for European agriculture and rural development. Aldershot (UK): Ashgate Publishing Limited.

RER (2007). Fattorie Didattiche dell'Emilia-Romagna – Guida ai percorsi didattici nelle aziende agricole. Bologna: Regione Emilia-Romagna.

Salghetti, A., Ferri, G., and Manghi, E. (2007). Strategie d'impresa e multifunzionalità in agricoltura. Conferenza Annuale Facoltà di Medicina Veterinaria di Parma, XXVII, retrieved October 31, 2009 from http://www.unipr.it/arpa/facvet/annali /2007/07_265_288.pdf.

Stinga, A. (2004). Galleria di ritratti. Il Divulgatore 9/10, 36-57.

Wills, W. J., Backett-Milburn, K., Lawton, J., MacKinnon, D., and Roberts, E. M. (2008). Parents' and teenagers' conceptions of diet, weight and health: Does class matter? Full Research Report ESRC End of Award Report, RES-000-23-1504. Swindon: Economic and Social Research Council (ESRC).

Food and tourism: the role of the "Slow Food" association

Stefano Buiatti

Department of Food Science
University of Udine, Italy
stefano.buiatti@uniud.it

Abstract

Tourism is an important economic activity in many countries, especially in Italy where the tourism industry is one of the major economic activities in terms of turnover. It is necessary that the traditional tourism activities, in particular those specialized in food tourism, develop a sensitivity to protect and integrate with the natural and cultural resources available in the territory. This paper analyses the role of intermediaries to create such awareness by focusing on the case of "Slow Food". Through the "tools" *Ark of Taste* and *Presidia,* the "Slow Food" association aims to protect food heritage, saving endangered food species through the "Slow Food Foundation of Biodiversity". Through food tourism it is important to generate economic development that can help preserve natural resources and improve the quality of life of the communities in a region. This ensures that indigenous food and production methods are not lost in the face of globalisation.

Zusammenfassung

Der Tourismus ist ein wichtiger Wirtschaftsfaktor in vielen Ländern, vor allem in Italien wo die Tourismusindustrie einen der wichtigsten Wirtschaftszweige insgesamt darstellt. Es ist notwendig, dass die traditionellen Anbieter touristischer Aktivitäten, insbesondere diejenigen, die sich auf den Gastronomietourismus spezialisiert haben, eine Sensibilität für den Schutz der natürlichen und kulturellen Ressourcen der jeweiligen Region entwickeln. In diesem Beitrag wird am Beispiel von „Slow Food" die Rolle von intermediären Institutionen bei der Erzeugung einer solchen Sensibilität analysiert. Durch die Instrumente "Ark of taste" und „Presidia" versucht die Vereinigung Slow Food mit Hilfe der „Slow Food Stiftung für Biodiversität" Ernährungs-Traditionen zu schützen und bedrohte essbare Tier- und Pflanzenarten zu erhalten. Es ist wichtig, die Gastronomiebranche zu nutzen, um wirtschaftliche Entwicklung zu erzeugen, die zur Erhaltung der natürlichen Ressourcen und zur Verbesserung der Lebensqualität der Gemeinden in der Region beitragen kann. So kann sicher gestellt werden, dass die einheimischen Lebensmittel und Produktionsmethoden angesichts der Globalisierung nicht verloren gehen.

Riassunto

Il turismo è una importante attività economica in molti paesi e in particolare in Italia, dove questa industria presenta un fatturato molto elevato. Relativamente al turismo gastronomico è fondamentale che questo possa diffondersi integrandosi con la protezione e valorizzazione delle risorse naturali. Questo lavoro valuta il ruolo di soggetti intermediari nel sostenere questo tipo di sensibilità, prendendo in esame il caso di "Slow Food". L'associazione Slow Food, grazie all'Arca del gusto e ai Presidi, ha come scopo quello di difendere le produzioni locali e

K.L. Sidali et al. (eds.), *Food, Agri-Culture and Tourism: Linking local gastronomy and rural tourism: interdisciplinary perspectives*, DOI 10.1007/978-3-642-11361-1_6,

tradizionali, salvare le specialità alimentari e agricole difendendo la biodiversità grazie alla Fondazione Slow Food. Il turismo gastronomico è fondamentale per favorire uno sviluppo economico che aiuti a preservare le risorse naturali e ambientali, aiutando a migliorare la qualità della vita delle comunità coinvolte. Questo consente di offrire maggiori garanzie a quelle specialità alimentari e metodi di produzione che rischierebbero di andare perduti a causa della globalizzazione.

Keywords: Slow Food; Tourism; Food tourism.

1 Introduction

Slow Food is a non-profit, member-supported organization founded in 1989 to counteract fast food and fast life, the disappearance of local food traditions and people's lack of interest in the food they eat, where it comes from, how it tastes and how our food choices can affect the economy of the rest of the world.

Slow Food began in Italy with the foundation of its forerunner organization, Arcigola, in 1986 to resist the opening of a McDonald's near the Spanish Steps in Rome. The Slow Food organization spawned by the movement has expanded to include over 100,000 members with chapters in over 132 countries. In total there are about 800 local *convivia* chapters. There are 360 *convivia* in Italy (known as *condotta*) comprising 35,000 members, along with 450 other regional chapters around the world. The organizational structure is decentralized: each *convivium* has a leader who is responsible for promoting local artisans, local farmers, and local flavors through regional events such as taste workshops, wine and food tastings, and farmers' markets. Opposed to the culture of fast food, the Slow Food association wants to encourage the enjoyment of regional products and traditional foods.

2 From the pleasure of food to food tourism

Food is a common language and, above all, a universal right. As any language, it is spoken to communicate, to share emotions, feelings, sensations. The pleasure of food is not only based on the taste but on the sharing of it with others. Nobody can really enjoy their food without considering that this is a universal right and that each kind of food, even the most common and "simple", presents many histories that could be told, history of a region, history of an identity, history of a population, history of a village, history of a family, history of a religion, in a word, history of a culture.

Sharing food means talking about the pleasures of the table and this, in turn means talking about hospitality. In many languages which have their root in ancient Latin, the word "*hospes*" has a double meaning. The same word designates both the person who welcomes and the one who is welcomed, host and guest. The anthropological and sociological meaning of that might be that the guest has the same rights and duties of the host and this goes beyond the old (think of the pilgrims during the middle ages) and "asymmetrical" idea of the relationship between an authoritative active role (the host) and a fatally and inevitably passive one (the guest). Only bearing in mind this concept it is possible to talk of high quality and modern tourism.

How can a territory, a region, a village and their food specialties, be linked to tourism? Food is an important aspect of a traveller's experience. Most potential visitors to Italy want to taste traditional dishes, regional specialities, and fresh local produce, possibly organically grown. Food is an important economic and cultural resource offering tangible benefits for tourism. However it is essential that people involved in this activity are effective in delivering an authentic food experience to an increasingly aware and health-conscious group of visitors. In the definition of food tourism it is necessary to differentiate between the tourist who consumes food as a part of the travel experience and the tourist whose interests, behaviours, and even destination choices are influenced and determined by a specific interest in food. Food is obviously an integral part of any travel experience but its importance and significance has been ignored for different reasons: for some travellers it is just a kind of "fuel", for others, who are a growing minority, it is a significant, at times the most important, reason or motivation to travel. Somewhere in between we can find all potential consumers/tourists who might become the driving force of food tourism.

The crowd of potential food tourists are becoming increasingly interested in this topic, and several indicators can confirm this: Of course a tourist is, first of all, a consumer, and from the consumer perspective the concern about the quality and safety of food he/she can find on shelves and in restaurants becomes stronger every day, especially after the BSE (Bovine Spongiform Encephalopathy) crisis, the foot and mouth outbreak in 2001 and the more recent avian or "bird" flu. Also very important is the recognition by "green" consumers of the impact that long distance food transport can have on the environment (the so called "food miles"). This has led to a growing interest in organic and natural products, and the importance of food traceability and, hence, an increasing demand for more locally sourced food. The preservation of the countryside as a source of relaxation and rest is a fundamental issue both for the tourist/consumer and the tour operators that benefit from it. Also important is to regard the tourism and the maintenance of a landscape as an important potential solution to sustain the social life of people that live there, creating productive activities and employment.

Food tourism can be, and in many cases already is, a driving force for farmers to rethink their production activity taking into account the diversification into new markets and the opportunities to obtain an added value to their products. This is in line with the philosophy of Slow Food, which is to achieve long-term change in the areas of biodiversity, preserving threatened foods and strengthening local economies through assisting small-scale producers and building short-supply chains; taste education, creating a better understanding and appreciation of the important connection between health, local food culture and agriculture; and finally, food communities, connecting producers to consumers, creating awareness that showcases sustainable agriculture and artisan food production.

"According to this philosophy the food should taste good: excessive food miles, standardisation and homogeneity very often result in bland food that simply doesn't taste as good as food that has been produced locally with care, pride and passion. The food should be clean, produced in a sustainable way that does not harm the environment, animal welfare or our health, and with a minimum of artificial intervention. And finally the food should be produced in a fair way: its producers should not be exploited, but should receive fair compensation for their work and skill" (retrieved January 10, 2010, from www.slowfoodfoundation.org). According to these three fundamental issues Slow Food considers all people co-producers and not only consumers, because by being informed about how our food is produced and actively supporting those who produce it, everybody becomes a part and partner in the production process.

Food tourism is strictly related to the literature on wine tourism which is defined as a tour or visit to wineries, wine festivals and events correlated where the most important motivation for the tourist is to directly experience the characteristic, the quality of a wine through a tasting session. In the same way food tourism could be defined as 'a visit to food producers with a strong interest in tasting that food and an even stronger motivation to learn about the production process from raw materials to the final product.' The concept of food and wine tourism can be linked to the land and when, for example, the viticulture is successful it "changes" the local landscape into a mix of tourism, agriculture and industry. Consequently, food/wine production and tourism should rely on regional or local labels/branding for market promotion and in this way the regional/local brand can be an important source of differentiation and a sort of added value for a rural region where that particular wine or food specialty is produced. The topic of food specialties is of great interest nowadays and can be considered from different points of view, in particular the relationship between products and territory and the connection between agriculture and tourism. Food is an important part of regional culture and identity, and food production has a strong impact on the landscape.

3 The promotion strategy of Slow Food: Presidia and the Ark of taste

Basically, the Slow Food association has two tools to promote and protect typical foods: *presidia* projects and Ark of taste. The *presidia* began in Italy in 1999 as the supporting arm of the Ark of Taste. The Ark had catalogued hundreds of products at risk of disappearing, but with the *presidia*, Slow Food decided to make a concrete contribution to the world of food production. *Presidia* are local projects that focus on a group of producers of a single product, developing production and marketing techniques to allow their work to be economically viable. The *presidia* program is the tool that Slow Food uses to assist producers directly in the commercialization, protection and promotion of their products. The *presidia* sustain quality production at risk of extinction, protect unique regions and ecosystems, recover traditional processing methods, and safeguard native breeds and local plant varieties.

The *presidia* directly involve producers, offer technical assistance to improve production quality, organize exchanges among different countries, and provide new market outlets (both locally and internationally). There are nearly 200 *presidia* in Italy, protecting the most various products: from Resia garlic to Garfagnana Potato Bread. With more than 120 international *presidia*, Slow Food protects biodiversity in the whole world: from Rimbas Black Pepper in Malaysia, to Mananara Vanilla in Madagascar. The Ark of Taste aims to rediscover, catalogue, describe and publicize forgotten flavours. It is a metaphorical recipient of excellent gastronomic products that are threatened by industrial standardization, hygiene laws, the regulations of large-scale distribution and environmental damage.

Ark products range from the Italian Valchiavenna goat to the American Navajo-Churro sheep, from the last indigenous Irish cattle breed, the Kerry, to a unique variety of Greek fava beans grown only on the island of Santorini. All are endangered products that have real economic viability and commercial potential. Thanks to the contribution of experts and collaborators around the world, the Ark currently includes more than 500 products from 30 countries. Developed in Italy in 1996, the Ark of Taste project predates the presidia project, and in many ways it is the 'grandfather' of the presidia. For many *presidia* products, inclusion in the Ark of Taste is the first step in project development, although this is not always the case.

4 The objectives of the presidia

The general objectives of the presidia cover the following four areas: economic environmental, social and cultural. The former are the only ones that can be evaluated using quantitative indicators such as variations in price, quantities produced, numbers employed, etc. The other objectives require a different approach and are more difficult to classify and compare using standard parameters, but they are perhaps the most interesting and important dimension of presidium activity. In the following an overview of the main objectives of Slow Food association is presented:

Economic objectives (increasing producer incomes, developing local driven activities, increasing employment). This aim is obviously extremely important. In fact, presidium products were disappearing because they were no longer profitable and in order to continue their activity, producers must have economic assurance about the future.

Environmental objectives (defending biodiversity, improving the sustainability of products) are essential for all *presidia.* Every set of production rules requires producers to eliminate or reduce chemical treatments, guarantee animal wellbeing (using extensive farming systems, adequate space, no forced feeding), defend, where possible, local breeds and native vegetable varieties, use ecological packaging, and to favour the use of renewable energy.

Social objectives (improve the social position of producers, strengthen their organizational capacity) can be measured by seeing whether the presidium has created an association or other organized body, if producers have improved their ability to form relationships with public and private institutions, if they are more widely known and if their voice has greater weight, thanks to media attention.

Cultural objectives (strengthening producers' cultural identity and promoting production areas) are linked to the ability of the presidium to stimulate publications about the local area, the creation of tourist routes and other cultural initiatives, the restoration of old buildings, etc. Considering the latter aspect, the synergy between food and tourism is evident: food (in general terms, not necessarily as a *presidia* product) can promote tourism thanks to the promotion (knowledge) of a territory or a region or a village and vice versa.

Table 1 displays the main goals of the Slow Food association from the four areas listed above.

Table 1. Main goals of the Slow Food association

Main goals of Slow Food Association
To protect the heritage of food, local tradition and different cultures
To defend food biodiversity
To safeguard the environment and the land
To promote taste education in order to help people to make daily choices about food with awareness and responsibility
To endorse sustainable agriculture
To protect small producers and their communities
To promote the gastronomic traditions of the whole world

Source: retrieved January 7, 2010, from www.slowfood.com

The mutual benefit for tourism and increase in value of food is evident. Tourism provides opportunities for food producers and vice versa, thanks to several aspects: for a producer it is possible to improve its identity values linked to that region, and this can represent a stronger reason for the consumer/tourist to buy that particular product; the moment of the visit is a chance for the producer to have direct contact with the consumer/tourist and consequently, an important opportunity to show the real production process.

5 Food and development of rural areas

In the following, some examples of *presidia* products are presented in order to show how 'quality is made tangible and not only advertised'. In some cases the food is the driving force and allows for a greater knowledge of a region (or a village, etc.); an example can be represented by the famous wine Chianti which is not a species of vine. Actually the name comes from the Chianti region of Tuscany, in Italy, where the wine is primarily produced. In fact, as is well known, the wine itself is derived from a variety of wines: the two primary red grapes are Sangiovese and Canaiolo, the whites are Trebbiano and Malvasia.

A definitely less famous example, but very illustrative of this kind of driving force of food, is represented by a "humble" *presidia*, the garlic of Resia, a small village of less than 1300 inhabitants in the Northeast of Italy. Resia garlic is a local ecotype, selected over the centuries in the Resia Valley in the Julian Prealps. As a result of its isolated position, the Resia Valley has managed to preserve a unique culture and language of early Slavic origin, as well as a remarkable range of plant biodiversity. The garlic has a medium-sized bulb with pink-streaked skin, while inside the white cloves are firmly attached in a circular formation. This arrangement of the cloves in a single ring is unusual, as most other varieties have additional internal rings of cloves. There was once a flourishing trade in Resia garlic, sold in markets as far off as Ljubljana and Vienna mainly because it is much more digestible than the ordinary garlic. Resia garlic has been a *presidia* for 5 years and since then as soon it is ready to be sold, there is a flow of visitors (consumers/tourists) because the harvest is rather limited, only five tons per year. Garlic is a very simple, plain vegetable; it does not have the nobility of other more renowned and famous gastronomic products of Italy. In spite of this it has an important social and economic impact on the village and the valley.

Another example comes again from the Friuli Venezia Giulia region (north east of Italy, near the Austrian and Slovenian borders) where the raw ham made S. Daniele (Udine) world famous for this typical product. A similar situation can be described for Parma, a city in the Emilia Romagna region also very famous for its raw ham. In other cases, tourism becomes the driving force that can allow the tourist, who is obviously also a consumer, to get to know directly a typical product or products of a particular region (or a city, village) already well known for its natural beauties; a good example is Venice, obviously known in the world not for its food specialties but nowadays very appreciated for its gastronomic offers, too. *Bresaola*, the air-cured beef of Valtellina, represents a further example: Valtellina is a touristic area of Lombardy region, very famous both for summer and winter holidays. *Bresaola* became famous (and is now exported to the USA too) thanks to the flow of visitors in this valley.

Surely the strongest example is represented by the wine routes, nowadays very common in many Italian regions, the most famous one in Tuscany thanks to Chianti wine, as already described. Italy is a country with rich gastronomic traditions and typical products to discover while following interesting food and wine routes. Wine tourism, developed as an economic activity in the last decade, is not something new. In fact the wine culture and the discovery of regions, areas and routes originated in countries such as Germany and France, with route projects aimed at promoting direct sales to the consumer/tourist. This kind of situation has existed for decades, yet it is comparatively new for Italy, in spite of a long tradition linked to wine culture in each region of Italy, from north to south, islands included. Wine tourism can be seen as a tool to develop tourism in deprived areas and as valorisation of agricultural products and their territories. To promote and regulate the wine tourism, Italy promulgated a law in July 1999: No. 268. The interest in oeno-gastronomical tourism has increased greatly in recent years, and Italy offers food and wine routes in many regions. Italian oeno-gastronomical tours are planned in order to promote the appreciation of Italian produce and its land. In many regions, tourists can visit the main agritourism farms of the area and taste typical products, as well as visiting the monuments and artistic sites in the cities.

The importance of food and wine routes is still underestimated: for some regions, wine routes can represent a sort of surplus in addition to an already very rich tourist offering in terms of traditional tours (history, art, monuments, etc.), for other regions, not famous at all in terms of tourism, they can become a fundamental tool to stimulate the local economy. Indeed wine routes allow the tourist/consumer to learn a lot about a territory and to discover the traditional tourist offerings of a region (for example seaside resorts in summer or skiing holidays in winter). This is the case for the Italian region of Friuli Venezia Giulia which is close to the Slovenian border. In tourist terms, until recently this region was not very well-known compared to more famous Italian regions (Tuscany, Sicily, Umbria, Veneto, Sardinia, Liguria, Trentino Alto Adige, etc.). In fact, in the past, the only relevant tourist resort often visited was Aquileia, a Roman archaeological site (during the imperial period Aquileia was the second largest city of the Roman empire) and two coaches arriving in the wine area of the region were enough to fill the hotels or Bed and Breakfast of that area.

Today the situation has changed dramatically and the tourism linked to food and wine production has become crucial for the regional economy. Accommodation in agritourist structures (farm holidays) or in wine estates has been strongly promoted and funded by the regional authority. Wine makers, restaurants and hotels have formed a regional branch of a national association, called *Movimento Turismo del Vino* (Wine Tourism Movement) based in the region of Umbria. "This is an association created in 1993 that has encouraged Italian wineries to open their doors to visitors: Italy has now become the country of "*Cantine Aperte*," ("Open Cellars") the annual event organised by the association on the last Sunday of May when wineries welcome visitors and wine lovers (consumers/tourists). It is a non-profit association with the aim of promoting wine tourism and improving the wine areas' image and prestige, as well as their economic development. Wine Tourism Movement wants to stand for the protection of the environment and for quality agriculture; promoting a natural life-style that puts the visitor at the centre of attention. Among the almost 900 members of the Association are some of Italy's most prestigious wineries, selected on the basis of very precise characteristics, the first of which are their high quality of visitor hospitality" (retrieved January 10, 2010, from www.movimento turismovino.it).

The Friuli Venezia Giulia wine production is rather limited (less than 1 million hectoliters) and represents only 2% of Italian wine output, but its quality is, on average, very high and it can be said that the winemaking sector is the driving force of agricultural economy of the region. Furthermore, the "Open Cellar" event paved the way to the discovery of other regional products, from meat products to dairy products, from fruits to vegetables and other food products such as honey and preserves.

6 Conclusions

Food and wine tourism is strictly related to the development of a sustainable tourism which has as its main objective to minimize environmental and cultural damage, maximize tourist satisfaction, and optimize long-term economic growth for the region. It is a way of obtaining a balance between the growth potential of tourism and the conservation needs of the environment. Moreover, "by participating in food festivals and events, by visiting and buying from specialist food factories, by buying from farm shops and farmers' market and by eating at restaurants, pubs and inns which feature local foods/dishes on their menu, the tourist connects closely with the local food culture, starts to understand the landscape that provided it and directly supports the rural economy" (Hall, 2003).

All these opportunities require human resources and intelligence, people informed and able to acquire new awareness, able to have a new model and a new ethic from which to see food tourism development and the natural resources as complementary for their respective sustainability. As Francois de La Rochefoucauld (1613-1680) wrote in XVII century it should be always remembered that "to eat is a necessity, but to eat intelligently is an art".

Acknowledgements

I sincerely thank Dr. Giulio Colomba, vice president of the "Slow Food" association, for his kind help and availability.

References

Antonioli Corigliano, M. (1999). *Strade del vino ed enoturismo. Distretti turistici e vie di comunicazione*. Milano: Franco Angeli.

Hall, C.M., (2003). *Food tourism around the world: development, management and markets.* Oxford.

Italian law No. 268; July 27, 1999, Disciplina delle "Strade del vino", Gazzetta Ufficiale della Repubblica Italiana, No.185, 9 agosto 1999.

OECD (1999). *Cultivating Rural Amenities. An economic Development Perspective.* Paris: author.

Opportunities of cooperative marketing using the example of the wine region Saale-Unstrut

Axel Dreyer
Juliane Müller

Tourism Marketing
University of Applied Sciences of Harz
prof.dreyer@t-online.de
juliane_mueller1@gmx.de

Abstract

This article shows which possibilities tourism and viniculture have for a better commercialisation of their products. It points out how the economic welfare of a whole region can benefit from the close cooperation between these two trade groups. Based on a mutual relationship between viniculture and tourism, various forms of cooperation are explained. Subsequently, marketing measures arising from this cooperation are presented both from the perspective of tourism service providers as well as vineyard owners. Product combinations of wine and other main attractions (sightseeing at cultural points of interest, cycling, and so forth) in the Saale-Unstrut-Region - one of 13 German wine-growing regions - are presented as an example to round off the article.

Zusammenfassung

Der Beitrag zeigt auf, welche Möglichkeiten Tourismus- und Weinbranche haben, um ihre Produkte besser zu vermarkten. Es wird verdeutlicht, dass mit Hilfe enger Zusammenarbeit beider Seiten das Wohlergehen einer gesamten Region profitieren kann. Ausgehend von den Zusammenhängen zwischen Weinbau und Tourismus werden verschiedene Formen der Kooperation erklärt. Anschließend werden die durch Zusammenarbeit entstehenden Vermarktungsmöglichkeiten sowohl aus dem Blickwinkel der touristischen Leistungsträger als auch der Weinbaubetriebe aufgezeigt. Am Beispiel der Saale-Unstrut-Region, einer von 13 deutschen Weinregionen, werden zur Abrundung Produktkombinationen von Wein und anderen touristischen Schwerpunkten (Besichtigung kultureller Sehenswürdigkeiten, Radfahren etc.) erläutert.

Riassunto

Il presente lavoro illustra quali possibilità hanno il turismo e la viticoltura per una migliore commercializzazione dei loro prodotti e sottolinea come un'intera regione può trarre beneficio dalla stretta collaborazione di queste due attività in termini di benessere economico. Vengono spiegate varie forme di collaborazione basate su un interscambio tra viticoltura e turismo. Successivamente si illustrano sistemi commerciali originati da tale cooperazione sia dal punto di vista degli operatori turistici che da quello dei viticoltori. La presente relazione si concentra sulla regione Saale- Unstrut, una delle 13 regioni vitivinicole della Germania, che presenta in modo esemplare alcune possibilità di abbinamento tra il vino e le altre principali attrazioni turistiche (visite ai luoghi d'interesse artistico-culturale, sport ciclistico e così via).

Keywords: Wine tourism; Cooperation; Destination marketing.

K.L. Sidali et al. (eds.), *Food, Agri-Culture and Tourism: Linking local gastronomy and rural tourism: interdisciplinary perspectives*, DOI 10.1007/978-3-642-11361-1_7,

1 Connections between viniculture and tourism

Viniculture and tourism can benefit from each other. Vineyards are predominant in landscapes that are of interest for tourists generally interested in vineyard destinations. Landscapes with vineyards are – not only in the perception of tourists – an elementary part of the tourism potential of a region as the following chart show.

Figure 1. Importance of wine landscape

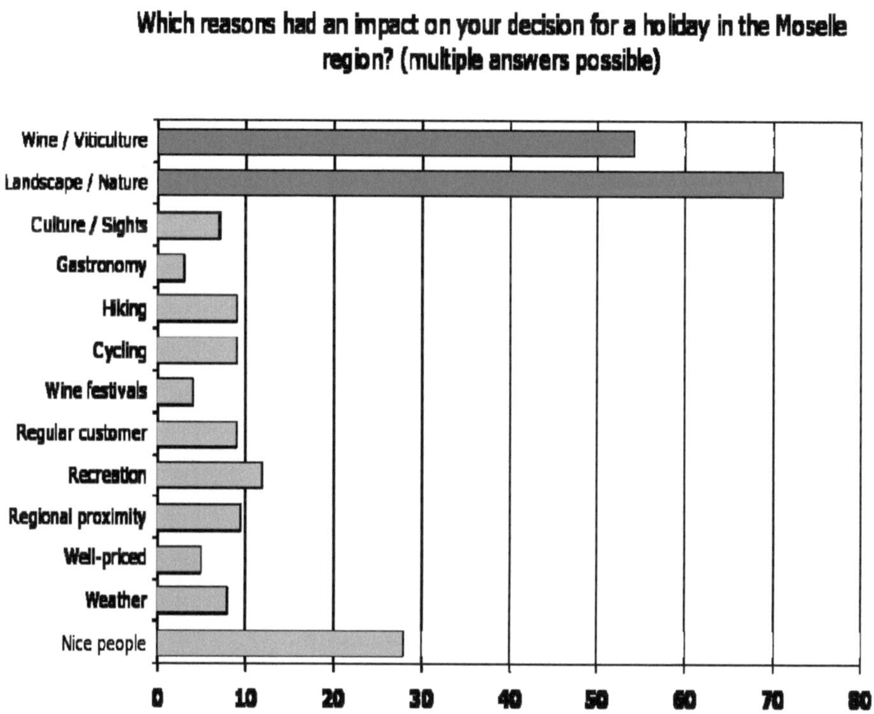

Source: Haart (2007)

From a tourist's point of view, the landscape is mainly natural scenery. Apart from national parks, tourist areas are shaped and changed according to the requirements of tourists. Tourist landscapes are "properly" arranged for tourists i.e., information signs, wooden benches, rubbish bins and so forth are set up in a landscape where they originally would not be found. The target is to arrange the landscape sites as the guests wish them to be. For successful tourism marketing only in very rare cases a wonderful landscape or intact nature can solely be taken into consideration. Without special landscape highlights a tourism offer can only be one amongst others.

Figure 2. Gosecker Dechantenberg – typical scenery of the Saale-Unstrut wine region

Source: Dreyer

Viniculture, in particular vines on steep slopes deliver a unique picture of a landscape with which a region can position itself thematically and can highlight itself from other competitive regions which only advertise beautiful sceneries. The picture of the landscape as a frame and base for a tourist offer is preserved, extended, and cultivated by the wine-growers and wine-makers through their production activities.

Tourism brings customers to the wine-growers which enable a direct sale of their products and thus better product profitability than sales via retailers. By means of tourism offers like wine festivals, wine-cellar visits, or 'Straußwirtschaften' (seasonal wine taverns) the temporary restaurants also offer the possibility to realize a not inconsiderable additional income. Tasting wine for instance often leads to buying the tasted wines.

This symbiosis is promoted particularly on the part of wine makers rendering services to tourists. Guests in a wine region expect the highest possible authenticity and that in particular is what wine makers provide. Overnight accommodation offers like 'staying with the wine maker', are very popular in this respect.

Figure 3. Symbiosis of wine makers and tourism industry

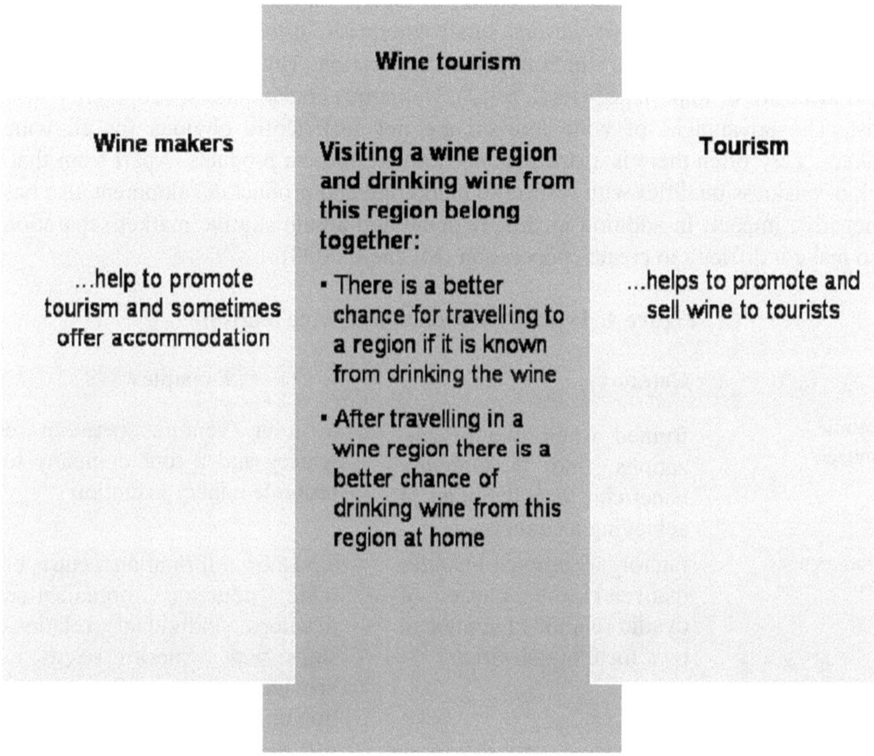

Source: own representation

2 Forms of cooperation in the wine tourism sector

In order to be able to optimally organize wine and tourism marketing all possibilities from the point of view of the tourist businesses and tourism organisations as well as the position of the wine makers and the respective wine growing organisations must be considered. Furthermore, each party must gain an insight into the situation of the others concerned. It is only possible for both sides to fully benefit from the synergistic effect of merchandising joint marketing activities if both parties mutually understand each other's performance and individual marketing activities.

Various forms of cooperation and networks are possible in the field of wine tourism. According to Hall et al. (2008) partners of all sizes can be included. "They can be locally or internationally based, they can occur at all stages of the value chain, and they range from highly informal relationships through to contractual obligations."

(ibid., p. 206). They also point out that cooperation in the field of wine tourism can be difficult to organise. Relationships between numerous different partners are necessary, partners which often consider themselves a part of a separate industry and pursue different business purposes (ibid., p. 208).

Getz (2000) also alludes to the difficulties of creating a network in the field of wine tourism. A large number of various small enterprises often do not have sufficient comprehensive knowledge, which impedes cooperation. Thus, national, regional, and local ideas are of importance (ibid., p. 12). Numerous constraints on cooperation also exist. The advantages of wine tourism are not sufficiently obvious for all wine makers. Very often there is a strong concentration on own products. Apart from that, lack of business qualities with respect to marketing and product development also has a negative impact. In addition to this, regional and administrative market separation also make it difficult to create cooperation (Mitchell, 2008).

Figure 4. Types of cooperation of wine tourism

Category		Example
Dyadic linkage	- formed when two organizations find it mutually beneficial to collaborate in achieving a common goal	- a joint venture between a winery and a tour company to promote winery visitation
Organization sets	- interorganizational linkages that refers to the clusters of dyadic relations maintained by a focal organization	- a visitor information centre or wine tourism organization develops individual relationships with wineries so as to provide tourists with information on each winery
Action sets	- a coalition of interacting organizations that work together in order to achieve a specific purpose	- a visitor information centre and the wineries in a region come together to produce a regional wine tourism promotional campaign
Networks	- used here in a narrow formal sense, refers to a group of organizations that share common organizational ties and can be recognized as a bounded interorganizational system	- a federation or association of wine tourism organization, e.g., the Movimento del Turismo del Vino; The European Council of Wine Regions; the Dyonisos multimedia network of European wine - producing regions

Source: Hall, Johnson, and Mitchell (2008, p. 208)

Antz, Dreyer, and Linne (2006) achieved similar results with their conceptual analysis of wine tourism in the Saale-Unstrut-Region. When interviewing local groups they had been told that cooperation "works quite well" but closer research revealed the contrary. One party did not know what the other was doing and there was even a lack of basic requirements for cooperation namely good communication and interest in discussions. In an interview for instance one expert said that "in the region there are very few businesses or gastronomes who are interested in a mutual approach." (ibid., p. 206).

The advantages of cooperation in the field of wine tourism can be manifold. Carmichael (2001) stresses the possibility to bundle budgets and create a "regional image" (ibid., p. 247). With this respect, Walton (2007) adds the importance of creating a brand and strengthening the popularity of the wine region. "Tourists prefer to travel to a wine region especially if they have the possibility to visit various vineyards but nevertheless get the impression of an overall connection" (p. 24).

There are various possibilities for cooperation in the field of wine tourism. The so-called stakeholder networks include all those interested in wine tourism. At the same time, each individual has an important part to play. Cooperation between vineyards and tour operators is helpful for developing wine tourism in a region. Hotels and guesthouses are another group of important partners. Primarily they offer sufficient accommodation possibilities which are not available on a wine-growing estate or vineyard. Apart from that they also provide information for tourists concerning regional offers about wine tourism. Furthermore, cooperation involving tourist information centres, regional trade, restaurants, and public institutions is recommendable (Walton, 2007; p. 24). Cooperation involving vineyards and a destination management organisation is also an essential key for successful marketing in the field of wine tourism. One component in this respect is the joint use of customer data for marketing purposes (Getz, 2000; p. 203).

Figure 5. Saale-Unstrut-Wine as displayed in the Online-Shop of Jacques Weindepot

Source: Screenshot retrieved June 6, 2006, from: www.jacques.de

Vertical as well as lateral cooperation exist with respect to wine tourism. Horizontal alliances arise from a cooperation of vineyards amongst each other. This in particular has a positive effect as far as marketing is concerned. Financing brochures and other promotion material is easier if budgets are consolidated. Lateral cooperation occurs when vineyards cooperate on the same level with partners in the tourism sector (Rasch and Gretzel, 2008; p. 320).

There are only few economic branches that, like viniculture, are similarly suitable for cooperation in the field of tourism. One could even say that the wine making industry has a natural affinity to tourism. This for instance is also underlined by an offer in the online-shop of Jacques‘ Weindepot (see Figure 5) and also becomes obvious in the following chart (see Figure 6) which depicts possible synergy effects existing in the wine and tourism sectors.

Sales promoters for Jacques' Weindepot let the wine makers present their own wine. Short stories are told about the wine makers and the vineyards and informative material about wine is handed out. This adds an emotional component to the presented wines and increases the feeling of quality for the target groups. The example shows that a vertical cooperation can also be of interest for distributive trade. Another form of vertical cooperation can be sought in the gastronomy and catering sector.

Figure 6. Synergistic effects of wine and tourism promotion

Wine-promotion and -distribution through tourism business		**Tourism-promotion and -distribution through viticulture**	
- Tourist information centers	- Wine presentation and sales - Information about different wine offers in a region - List of wine makers to visit - List of opening hours of seasonal wine taverns ("Straußwirtschaften") - Providing customer data for wine promotion	- Wine maker /grape producers/ wine growers cooperative	- Providing wine-related offers for tourists - "Straußwirtschaften" as typical regional gastronomy - Accommodation offers like "staying with the wine maker" - Tourists working as fruit pickers (volunteer tourism) - Tourism advertising by information on the label and bottle tags - Providing customer data for tourism promotion
- Accommodation	- Information about different wine offers in a region - Special wine hotels or creation of wine-themed hotel rooms - Providing wine-related offers - Staff members as important multiplicators		
- Gastronomy	- Selling wine and information about regional wine makers - Organization of wine tastings - Staff members as important multiplicators - Regional wine lists with information about wine makers	- Viticulture association/ regional wine marketing organization with touristic accents	- Information about and advertising of the wine region and the visitation of wine makers as tourist attractions
- Incoming-Agencies	- Finding innovative and attractive offers - Making packages for travellers - Providing the tourists as customers for wine makers	- Natural park	- Arranging wine sponsorships - Guided vineyard tours
- Tour operators	- Making packages for travellers to wine destinations including visits of vineyards		

Source: own elaboration

3 Cooperative marketing and wine tourism

The following chart shows by means of various examples how cooperative marketing can be realised. On the left side of the table the focus is on wine promoting activities of tourism service providers whilst the right side depicts advertising possibilities of the wine industry.

In order to be able to make optimal use of synergies, active marketing measures with reference to tourism and viniculture in a wine region should be centralised and controlled by one organisation and from one location. This means that promotion for wine in a certain region of one viniculture or wine makers' association and the organisation for tourism destination management should be unified in one organisation (with two departments). In consequence, the cooperating associations should be merged at one location and in one building. With such a procedure a wine region would achieve a decisive competitive advantage because this makes it easier to configure marketing activities more efficiently, which is not possible in other regions due to a structure involving multiple associations. In any case there is no joint wine and tourism organisation in Germany. So far only the viniculture association in Saxony and the tourism organisation in the Saxon-Elbe-Countryside-Region (Sächsisches Elbland) are located in one building and in the wine region Ahr promotion for wine and tourism is managed by one organisation the 'Ahr Rhein Eifel Tourismus and Service GmbH'.

The advantages of a joint organisation are obvious. Offices in one building are good for communication and minimize the loss of information. With a 'house of wine and tourism' it would be possible to increase efficiency in administration and for promotion activities. Some examples for this are:

- no duplication of advertising in brochures,
- a common data pool for direct marketing,
- one common homepage for the wine region,
- new ideas for joint advertising and so forth.

As far as intercompany activities are concerned increasing cooperation would also be an advantage for overall marketing activities in the region. Promotion measures based on cooperation between wine hotels and vineyards with lodging possibilities would unfold the full potential of marketing power and signalise: here is a rich offer of accommodation for tourists interested in wine and the wine region.

A closing of ranks between wine makers of a region as well as manufacturers of regional products and the gastronomy industry is of crucial importance, as many guests in particular look for regionally typical attractions. The range of products offered includes the respective service. This means that service staff in a restaurant should for instance be able to provide all the relevant information on local wine. These facts are also key arguments for why other trade associations like the hotel and restaurant association (in Germany DEHOGA) as an independent organisation should be located under one roof in a 'house for wine and tourism'.

Wine can be used as the strategic leading issue for tourist marketing and regional promotion, but the subject of wine should be combined with other issues in order to appeal to as many different target groups as possible, and in this respect a wine region in particular offers numerous possibilities. The unique selling proposition (USP) of a wine region is exactly that it is based on the many networking possibilities with wine and wine related activities and products. Apart from that, wine also has a positive image which can be transferred to the image of the travel region. Germany is a good example of this: interesting sights can be found almost everywhere but wine is only predominant in 13 regions. Sights can additionally make a wine region more attractive for wine tourists. Apart from that wine trips tend to be longer than visits to cultural attractions. It is also easier to encourage wine tourists to stay overnight in the region.

4 The Saale-Unstrut-region as an example for tourism product combinations with wine

Natural landscapes can be found in the Saale-Unstrut-Region: precipitous terraces located on centuries old dry walls, wine cellars in between orchards, wetlands, and biotopes with rare flora and fauna. Located high above this picturesque scenery are castles and fortresses. Culture and history, Romanesque and romanticism, nature and wine, here all this is quite naturally united. But the grape-vines give the riverside landscape its unique character.

Promotion activities function in such a way that the subject of wine serves as an emotional reason for the journey, and the uniqueness of the journey is established by the numerous cultural sights and the possibilities of an active arrangement of day activities.

Wine Region Saale-Unstrut =	Wine + Places of Interest
	or Wine + Activities

4.1 Wine and culture

Wine and culture fit together due to a high target group affinity. When taking into consideration age, income, and school education one can definitely see overlaps as far as the target groups are concerned, which are signs for positive synergetic effects with respect to a merger or bundling of marketing measures. That joint activities can increase wine consumption and also increase the utilization of cultural-historical offers has been successfully proven in various German wine regions. Viniculture itself offers sightseeing possibilities and presents a culinary frame for a successful journey.

Apart from wine culture in an actual sense (cultural landscapes characterised by viniculture or Germany's largest champagne producer „Rotkäppchen" with more than

100,000 visitors annually and so forth) the wine region with numerous historical sites as witnesses of the "Romanesque Road" ("Straße der Romanik"), places of "Sky Path" ("Himmelswege") and "Garden Dreams" ("Gartenträume") can offer a large variety of major sightseeing attractions linked to wine and culture within a small geographic area.

Figure 7. Castle Neuenburg as a part of the viniculture scenery

Source: Dreyer

4.2 The Romanesque road (Straße der Romanik)

The Romanesque Road leads tourists along a route of more than 1000 kilometres through the region of Saxony-Anhalt. Along this route the traveller passes a total of 80 important historical sites and experiences a historical insight into the Romanesque and Ottonian era. The high density of impressive buildings and the good condition of Romanic works of art is unique for this region and can otherwise only be found in Germany in the Rhineland, an aspect which makes the federal state Saxony-Anhalt especially attractive.

During the Middle Ages, several Roman monuments and buildings were also built nearby the rivers Saale and Unstrut. Today the wine region is one of the major tourism destinations along the first tourist route in Germany's eastern states. Significant sights in the region are amongst others the castle Neuenburg zu Freyburg, the monastery in Memleben, the cathedral in Naumburg, Bad Kösen with the Castles Rudelsburg and Saaleck, the Romanic House, the Cistercian monastry Schulpforta (with a vinotheque in the winery), as well as the castle church in Zeitz.

4.3 Sky path (Himmelswege)

Recent archaeological findings and excavations, in particular the finding of the so-called "Nebra Sky Disc", are the reasons for launching a project which aims at making important sites and ground monuments in the Saale-Unstrut-Region accessible and interlinking the various sites. The focus of the project "Sky Path" is on the "Nebra Sky Disc", which was discovered in 1999 on the Mittelberg hill and since 2008 has been exhibited in the museum for prehistory in Halle, the first tourist destination in the project. Other tourist destinations located in this wine region are the site of discovery of the "Nebra Sky Disc" on the Mittelberg hill near Wangen with the experience centre "Arche Nebra" which attracts over 120,000 visitors annually, an observation tower, and the 7000 year old reconstructed sun observatory in Goseck.

4.4 Garden dreams (Gartenträume)

Apart from the Romanesque Road project, another project devoted to both nature conservation and tourism called "Garden Dreams" was created to revive Saxony-Anhalt's garden history. From over 1000 gardens and parks the 40 most beautiful and historic sites were chosen to be shown to the wider public. Amongst others in the wine region Saale/Unstrut is the castle garden Burgscheidungen located at the river Unstrut and the castle garden Zeitz (site of the State Garden Exhibition in 2004)

5 Wine and active tourism

Apart from culture, active tourism is also growing increasingly popular in Germany and is well represented in this region. It is therefore an appropriate measure to combine wine with various forms of outdoor activities.

5.1 Wine and water touring (Blue Ribbon – Blaues Band)

All wine growing regions in Germany are alongside rivers and therefore water and wine are an ideal combination (slogan in offers: "water and wine"). In the wine region Saale-Unstrut are the rivers Unstrut, Saale, and the river Ilm near Bad Sulza. As far as the rivers Saale and Unstrut are concerned they already have a broad range of water related activities for tourists (canoe trips, rafting, excursion boats), which are promoted by the state tourism marketing association under the umbrella brand name "Blaues Band / Blue Ribbon". Various private service providers offer boat rentals or guided tours. The range of offered tours cover general sightseeing tours, visits to special attractions, or organised wine tasting events. The infrastructure (piers, sanitary facilities, sign-posting) seems to be well-developed. In particular, the upper course of the river Unstrut from Freyburg up to Memleben is considered placid and safe and can thus even be used by less sporty and ambitious target groups. At the confluence of the rivers Unstrut and Saale the infrastructure for water related tourist activities is already well-developed. At the "Blütengrund" (Blossom Ground) there is a pier for Unstrut river tours and a well adapted beer garden, there are piers for people interested in water sports, there is a sport equipment rental station, and a spacious camp site. The special combination of the proximity of small vineyards and the regionally typical wine cellars is another striking feature of this wine region. A

"Straußwirtschaft" (a seasonal wine tavern) in the historical vineyard of the painter and sculptor Max Klinger is only a few metres away. The Saale cycle tour route leads right past this location.

Figure 8. Lock operations on the Unstrut River 1992 – 2004

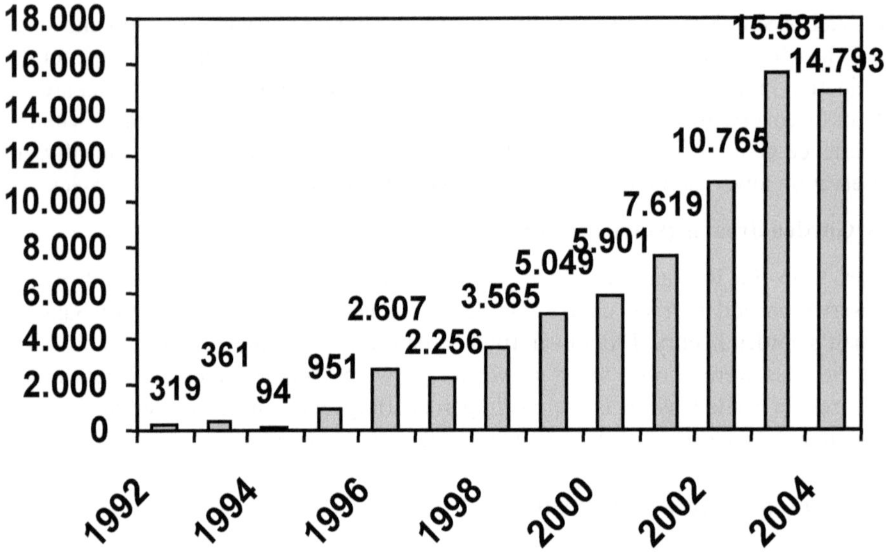

Source: Landesbetrieb für Hochwasserschutz und Wasserwirtschaft Sachsen-Anhalt, retrieved November 10, 2009, from http://www.sachsen-anhalt.de/LPSA/index.php?id=13427.

In interviews with experts, the increasing importance of water related tourist activities has been continually emphasised. Statistical data referring to the number of lock operations on the rivers prove an increasing high demand for water related tourist activities and sports of the blue ribbon offers.

5.2 Wine and cycling

Alongside the rivers Saale and Unstrut is an extensive national network of cycling tracks which are all well-developed and well-marked. As they lead alongside the rivers they also count as part of the "Blue Ribbon" project. The cycling tracks are the Saale biking route, the Unstrut cycle tour, the Elster cycle tour, and the Ilm Valley cycle tour. Very often though these offers lack a direct reference to wine and wine related products. And the route often does not pass directly by the vineyards, but only alongside them. The "Saalhäuser" route which leads through vineyards directly to a "Straußwirtschaft" (a seasonal wine tavern) of a winery is an example of this.

Figure 9. Cyclists in the Saalhäuser vineyard

Source: Dreyer

5.3 Wine and hiking

Nowadays hiking tourism is becoming increasingly popular again (Dreyer, Menzel, and Endreß, 2010). Hikers as well as cyclists need a destination for their activities. Here the numerous cultural sites offer possibilities, as well as the gastronomy and catering trade which must provide attractive offers to convince guests to enter and stay. It is evident that wine as a typical product of the region plays an important role. The - unfortunately not perfectly sign-posted - Saale-Wine-Hiking-Route already exists but the general offer of wine-hiking-routes for walking tours is still worthy of improvement.

5.4 Wine and wellness

The expression wellness was created by the American physician Halber Dunn. He combined the two words "Wellbeing" and "Fitness". What is meant is to achieve a special condition with a high degree of physical well-being by considering the human being in dependence of the environment as a holistic entity consisting of body, soul, and spirit. There is no clear definition of wellness but the various interpretations of the meaning in the broadest sense all have something to do with physical health and subjective well-being. How close the wellness-offer is connected to wine depends on the detail of the product design. For instance, grapes can be part of a healthy and varied diet rich in vitamins. Vinotherapy is one of the new the new trends for recreation and relaxation. This includes treatments like oil massages, exfoliation, or sauna infusions using products which are all based on grape seeds. In particular, the health resorts and spas of traditional health tourism like the spa Bad Kösen where they plan to increase the range of wellness offers, and the spa Bad Sulza could increase their unique selling propositions if they develop more wine related wellness activities.

6 Conclusion

The following chart depicts a summary of possible synergy effects between the wine and tourism industry. It is important to increase the commitment of the tourism and wine industry to shared values and goals. Tourism professionals must be willing to learn more about viniculture and on the other hand wine makers must increase their understanding of tourism.

Figure 10. Synergy effects of viniculture and tourism

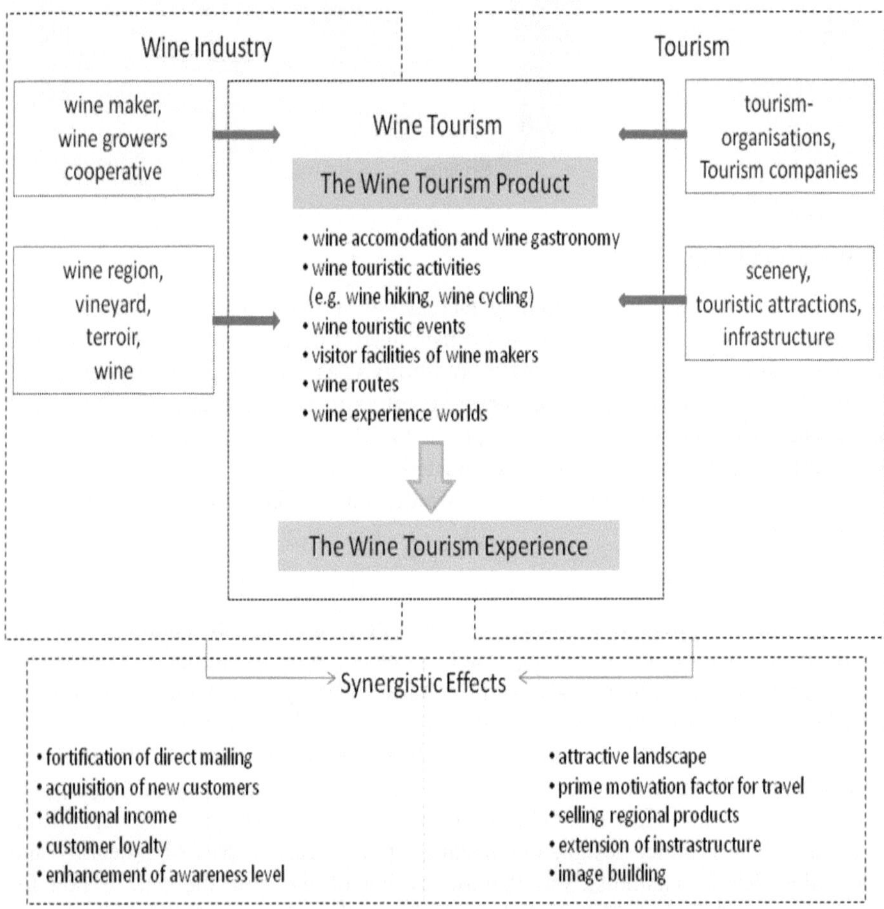

Source: own elaboration

In a "house of wine and tourism" the most important associations and organisations should be consolidated under the same umbrella with the target to improve communication between individual groups. Means of joint marketing activities are for instance wine fairs, road shows, or also a cooperation of wine hotels or vineyards with accommodation possibilities. It is important to promote the creation of these various forms of cooperation. In doing so, restaurants and catering businesses should also be included. It is essential to communicate to all interested people and the guests already enjoying the offers that it is an extraordinary experience to spend a few days in a wine region.

References

Antz, C., Dreyer, A., and Linne, M. (2006). Wein und Tourismus in der Weinregion Saale-Unstrut. In Ministerium für Wirtschaft und Arbeit Sachsen-Anhalt (Ed.), *Tourismus-Studien Sachsen Anhalt,* Vol. 25. Magdeburg: author.

Carmichael, B.A. (2001). *Competitive and Sustainable Wine Tourism Destinations, Travel and Tourism Research Association* retrieved October 11, 2008, from http://www.ttra.com/applications/DocumentLibraryManager/upload/Competitive%20and%20Sustainable%20Wine%20Tourism%20Destinations.pdf.

Dreyer, A., Menzel, A., and Endreß, M. (2010). *Wandertourismus. Kundengruppen, Destinationsmarketing, Gesundheitsaspekte*. München: Oldenbourg.

Getz, D. (2000). *Explore Wine Tourism: Management, Development and Destinations*, New York, Sidney, Tokio.

Haart, N. (2007). Weintourismus. In Becker, C., Hopfinger, H., and Steinecke, A. (Eds.), *Geographie der Freizeit und des Tourismus – Bilanz und Ausblick* (pp. 237-248). München: Oldenbourg.

Hall, C.M., Johnson, G., Cambourne, B., Macionis, N., Mitchell, R., and Sharples, L. (2008). Wine tourism: an introduction. In Hall, C.M., Sharples, L., Cambourne, B., and Macionis, N. (Eds.), *Wine Tourism around the world – Development, management and markets* (pp. 1-23). Oxford: Butterworth-Heinemann.

Hall, C.M., Johnson, G., and Mitchell, R. (2008). Wine tourism and regional development. In Hall, C.M., Sharples, L., Cambourne, B., and Macionis, N. (Eds.), *Wine Tourism around the world – Development, management and markets* (pp. 196-225). Oxford: Butterworth-Heinemann.

Mitchell, R., Hall, C.M., and McIntosh, A. (2008). Wine tourism and consumer behaviour. In Hall, C.M., Sharples, L., Cambourne, B., and Macionis, N. (Eds.), *Wine Tourism around the world – Development, management and markets* (pp. 115-135). Oxford: Butterworth-Heinemann.

Rasch, L., and Gretzel, U. (2008). Wineries' involvement in promoting tourism online: the case of Texas. *PASOS Revista de Turismo y Patrimonio Cultural, 6*(2), 317-326, retrieved November 11, 2008, from http://www.pasosonline.org/Publicados/6208special/PS0208_13.pdf.

Walton, L.A. B. (2007). Kentucky Wineries' Networking Activities for Tourism Development in Relation to Growth and Profitability. Dissertation, University of Kentucky, retrieved December 12, 2008, from http://archive.uky.edu/bitstream/10225/633/kywines.pdf.

Schätzel Vinery – Tradition 2.0

Kai Schätzel[a]
Ludwig Arens[b]
Jan Schätzel[a]

[a]Weingut Schätzel
Oberdorfstr. 34
55283 Nierstein
weingut@schaetzel.de

[b]Department of Agricultural Economics and Rural Development
Georg-August University of Göttingen
Platz der Göttinger Sieben 5
37073 Göttingen
larens@uni-goettingen.de

Abstract

The following case study describes how a vinery, by combining both tradition and new marketing and production concepts, has transformed itself from a sideline business to a major concern. In the last four years, the business manager has succeeded in repositioning the business and steering it towards growth, achieving a significant turnaround. Now, this 'young' traditional vinery must reinforce its position and establish concepts to ensure the long-term growth of the business. The measures employed to reach this goal will be described both in concept and in practice, so that they may serve as a best-practice model for the industry.

Zusammenfassung

In der folgenden Fallstudie wird dargestellt, wie ein Weingut durch Rückgriff auf Tradition sowie neue Marketing- und Produktionskonzepte von einem Nebenerwerbs- zu einem Vollerwerbsunternehmen umgestellt werden konnte. In den letzten vier Jahren ist es dem Geschäftsführer gelungen das Unternehmen neu zu positionieren und es auf Wachstum auszurichten. Der „turn around" ist gelungen. Nun muss das „junge" traditionelle Weingut seine Position behaupten und Konzepte zur Sicherstellung eines langfristigen Wachstums installieren. Die Maßnahmen zur Erreichung dieses Ziels werden im folgenden Beitrag konzeptuell und praxisnah beschrieben, sodass sie der Branche als best practice Vorlage dienen können.

Riassunto

Il presente articolo spiega come un'azienda vinicola, abbinando tradizione e nuove iniziative commerciali e di produzione, si è evoluta da attività secondaria ad un'impresa importante. Negli ultimi quattro anni la direzione è riuscita a ristrutturare l'attività e ad indirizzarla verso una crescita, completando un rilevante cambio di marcia. Adesso, questa "rinnovata", già tradizionale, azienda vinicola deve rafforzare la sua posizione e fissare le linee guida per assicurare una crescita duratura dei suoi affari. Indicheremo sia in teoria che in pratica i mezzi impiegati per raggiungere tale scopo, in modo che possano servire da modello teorico-pratico ottimale per altre aziende di questo settore.

Keywords: Viticulture; Quality management; Wine marketing.

K.L. Sidali et al. (eds.), *Food, Agri-Culture and Tourism: Linking local gastronomy and rural tourism: interdisciplinary perspectives*, DOI 10.1007/978-3-642-11361-1_8,

1 Introduction

Until 2006, the vinery of the Schätzel family had a marginal presence on the German wine market. It was not overseen full-time, which led to marketing losses, as the vinery was too large to be run part-time, but too small to support a full-time employee. The business was faced with the choice between ceasing production or using new concepts to expand. The decision to expand was made, and a full-time manager was found in the winemaker Kai Schätzel. In the last four years, the business has succeeded in repositioning itself and achieving substantial growth. Following this successful turn-around, this 'young' traditional vinery must reinforce its position and establish concepts to ensure the long-term growth of the business. The measures employed to reach this goal will be explained in the following article both in concept and in practice, so that they may serve as a best-practice model for the industry.

2 Historical overview

2.1 Business situation before the handover in 2007

Between 1977 and 2007 the vinery shrank from an initial 10 ha to about 4 ha of wine growing land as a result of the generally strained situation in the German wine industry during the 1980s. With the transition from a full- to a part-time business, the vinery lost momentum. Significant innovation and investment ceased, and the vinery was not able to secure a strong position in the constantly fluctuating wine industry. It was too small for cost-effective mass-production, yet lacked the innovative power to remain at the qualitative forefront of the market.

The production conditions remained largely unchanged in this compromise situation. Attempts to make the vinery operations more business-like achieved only partial success, since the vinery had barely changed since 1910. Despite reallocation of the vineyards, the majority of the vineyards in Nierstein remained scattered over large areas, and the location of some of the vineyards on steep slopes, make their management difficult. Although a modern grape entrance and machinery hall were installed, the wine cellars remained in a 600 year old vaulted cellar, where the wine was cultivated in old wooden barrels. Under these conditions, the vinery could neither produce cost-effectively in order to succeed in the price war, nor could the potential of the hillside vineyards and wooden-barrel aging be exploited to produce top-quality wines. Still under the influence of its previous manager, the vinery cast off its traditional look and designed new, very simple, labels. Between 1980 and 2000, the proportion of dry wines was increased to almost 95%. With this, the vinery, at that point under the name Schlamp-Schätzel, tried to release itself from the business structure of the Schlamp vinery. However, this lost them almost all of their business clients and marketing relationships, trying instead to concentrate on private customers and resulting in a drop in sales during this period from around 50,000 bottles (1984) to 22,000 (2005).

2.2 Business handover and reconceptualisation 2006 - 2009

In 2006, Kai Schätzel, the son of the previous owner, began to rebuild the business client sales in parallel to his studies. He subsequently committed himself full-time to the family vinery, where he worked until mid-2008. At this point, he established a BGB company with his father, and in 2009, ownership of the whole vinery was transferred to him. His mother remained with the business and concentrated on overseeing the private customer base and mail-order business. As manager, Kai organised the production and expanded the business client sales. Between 2006 and 2009, he instigated a radical change in strategy for the vinery.

2.3 Implementation and results up to 2010

The planned production reorientation and expansion was successfully implemented up to the end of 2009, with particularly the strict quality management showing rapid results.Numerous news items and strikingly good reviews in the specialist literature and wine guides pushed the 'young traditional establishment' back into the spotlight (Ali, Lecocq, and Visser, 2008). In a review of the best German wine brands in the specialist magazine *Vinum*, the vinery was placed at the top of the field (Scheuermann, 2009) and the renowned wine guide *Gault Millau 2010* considers the business a 'clear rising star' (Payne, 2010).

Particularly the concepts in business client sales were implemented more rapidly than planned. In this area alone, the vinery increased its sales between 2006 and 2009 by over 25,000 bottles, and thus more than doubled the original annual sales. The vinery's wines were introduced in numerous restaurants, including several renowned star restaurants, where they performed very well (Castriota and Delmastro, 2008). Their partner in the wholesale drinks industry could, as hoped, significantly expand their capacity in the wine industry. The difficult economic situation in 2008-2009 caused many in the food industry to rethink their situation, and careful matching of demand with supply from a one-stop drinks wholesaler was the right offer at the right time.

Overall, sales since 2006 have almost tripled, whilst at the same time the price rose on average 15 % per year. Despite increases in production costs, through the clearly more efficient utilisation of the business' resources, productivity was increased by over 20 %. Already in 2009, the business has generated profits and can make new investments from the current proceeds.

3 Overview of the designed strategies

The continually sinking sales to private customers between 1985 and 2005 brought the vinery in to serious difficulties. With 2005 marking their lowest sales turnover of 22,000 bottles, the business was barely viable. In response, a new business strategy was developed. Firstly, the core resources of the vinery were defined and relevant market segments identified. In order to reach these, a new product strategy was developed that led to a restructuring of production. To provide the link between production and the intended market segment, a new marketing and communications strategy was applied.

3.1 Core resources

As their most important resource, the business identified their exceptional vineyards in the best sites of the "Niersteiner Roten Hang". The combination of a unique soil composition (red clay-schist) and the steep south-eastern facing slope creates ideal conditions for the production of high quality Riesling wine. The vines are forced by the meagre soil to grow their roots up to 10 m deep into the ground in order to find sufficient water, the resulting high surface area of roots leading to an increased uptake of minerals. Additionally, the slopes expose the vines to a lot of sunlight, allowing the grapes to ripen particularly well. All of these factors make it possible to produce unique terroir wines from these vineyards (Hoffmann, 2009).

The vinery is endowed with a historical and attractive building in the middle of the town centre of Nierstein. The house stands over a 600 year old cellar containing traditional wooden barrels. The conservative development of the business after the Second World War meant that no plastic or stainless steel tanks were introduced, with which many other vineries replaced their old wooden barrels. Instead, the existing barrels were cared for, and the old vaulted cellar provides a perfect atmosphere to preserve and use these barrels, with care and attention, over many years.

It is recognised in today's production of top quality wines, that traditional aging in wooden barrels brings many advantages, which cannot be reproduced in stainless steel tanks (Schneider, 2005).

The historical buildings are surrounded by a large garden and vineyards. The business, although situated in the town centre, thus has enough space to further develop and expand.

3.2 Market segments

The logical consequence of the previously described core resources is the production of high quality, regional wines. In order to cover the necessary costs involved in this, a higher price segment must be targeted. In the private customer sector, this is the well-informed consumer, who should be made aware of the vinery through positive reports in the specialist media and recommendations (Ali et al., 2008; Castriota and Delmastro, 2008). In the business client sector, the vinery initially plans to concentrate on high class German restaurants. Through this, maximal breadth in variation of the abovementioned target groups of end users can be reached (Hoffmann, 2009).

3.3 Product strategy

The once very wide product range of well over 30 products was reduced to the significant core products. From the core resource vineyards in the "Niersteiner Roten Hang" it was decided that the focus of the vinery should be on the production of white wines. The favourite was the Riesling grape variety, best suited to convey the character of the soil all the way to the glass, followed by the Silvaner, for which the vinery had their own clonal variety that had been selected by Emil Schätzel at the beginning of the 19th century. In order to raise awareness in the targeted market segments, the top quality products from the best slopes were promoted at the top of the product pyramid ('General von Zastrow' label, cf. Figure 1). In the medium price segment, the character of the soil should be represented in its purest form ('ReinSchiefer' label). To conquer the 'by the glass' sales in restaurants, an additional plain and simple basic line ('ReinWeiß' label) was offered, which communicated the core message of the winery in a very simple manner and remains affordable. The sweet high-quality wines ('Heinrich Schlamp' label) round off the selection at the top end. The result is the product pyramid displayed below.

Figure 1. The quality pyramid of the Schätzel vinery

symbol	productbrand	classification	message	retailprice
	ReinWeiß	Rebsortenweine	one variety	5,50 - 6,50 €
	ReinSchiefer	Terroirweine	one terroir	7,50 - 9,50 €
	GeneralvonZastrow	Lagenweine	single vinyard	12,00 - 16,00 €
	HeinrichSchlamp	Prädikatsweine	off-dry	12,00 - 185,00 €

Source: www.schaetzel.de

3.4 Production strategy

In order to ensure the planned growth on the production side, a three pronged strategy was followed for short-, medium-, and long-term growth. The central idea for this strategy is that the business relies as much as possible on their own strength to grow, and to keep reliance on outside capital low.

The business considers it particularly important to manage the quality of the wine at the decisive points in the process themselves. Therefore the buying-in of finished wines on the barrelwine market is considered an option only in emergencies. In order to nevertheless be able to react flexibly to rising demand, the business pursues a 'middle-way' between the traditional renting of vineyards and the purchase of wine. Through contracts for buying in grapes from wine producers, the vinery plans to have more influence on the production and to be able to intervene at decisive points in the process. A reward system was developed which, in contrast to usual systems, does not remunerate the yield, but rather a fixed prices is paid per vine. In addition, the vinery does the treatment of the vineyards from July themselves in order to ensure high quality work there. It is thus possible to carry out quality-promoting reduction in the quantity of grapes without reducing the revenues of the grape producer. The grape harvest is always carried out by the vinery themselves, and the production of the wine occurs in their own cellars. This concept allows them to use high quality grapes without capital commitment and without relinquishing quality control to outside concerns.

In order to achieve even more influence over the cultivation quality in the medium term, the vinery plans to lease further areas of vineyard. This would allow the cultivation system of the vines and the methods of sustainable soil management to be even better attuned to the strategy of the vinery (Wörner, 1984). Those of their own vineyards that were rented out in the 1980s and 1990s will be reincorporated into their own production.

As soon as the commercial development allows, the vinery plans to buy further areas of vineyard in the best locations in Nierstein. The goal is that the areas for the production of top quality wine remain the property of the vinery long-term, and only these top products will be marketed as a "single vineyard wine" with the vineyard's own appellation. In this way, the product label will be built up (Gergaud and Livat, 2007). The risk of losing these locations again when grape purchase or lease contracts come to an end should also thereby be avoided. With the excellent locations they already own as well as a combination of short- and medium-term production expansion, a quality conscious course of growth is already being realised.

Also in the management, a concept stands in the foreground which emphasises not only quality enhancement but also growth with minimal capital commitment. The Schätzel vinery therefore decided to contract out the machinery work in the vineyards as far as possible to outside contractors, thereby avoiding cost intensive investment in tractors and equipment that a small vinery can poorly afford (Hanf and Schweickert, 2007).

Manual labour is an important prerequisite for the production of top quality wine, therefore an intensification of manual work under their own management was emphasised. Furthermore, running labour costs in the reorganisation and expansion phase of the vinery are easier to carry than large investments, which must be paid off over many years. The vinery is thus able to retain its flexibility to react to future challenges and yearly specific situations.

As essential quality measures, the vinery plans to carry out the cultivation, and therefore also the pruning, of the vines themselves. These measures are accompanied by an integrated approach to quantity reduction, whereby from May until September, shoots, grapes and part of grapes are continually removed in order to benefit the remaining grapes. The majority of the grape harvest should be carried out by hand, and between September and November, a staggered harvest should allow the selection of only the best grapes ready at that point. In order to recruit the appropriate labour for the above-described high quality work, an extremely flexible personnel concept is used. It is necessary to vary the body of staff between one and twenty, according to the season, and the number of permanent staff should stay as low as possible. However, at the same time, the demanding work requires competent personnel. The solution to the dilemma is seen in the employment of student workers, whereby it is possible to flexibly adjust the number of staff to the demands of the vinery. In addition, the vinery sees the students as interested co-workers that ideally build up an understanding of the finished product. A permanent member of staff coordinates and trains the student helpers.

In order to underpin the production of high quality wines with a sustainable approach, the vinery plans in the medium term to switch to organic cultivation (Gemmrich, 1998; Wucherpfennig, Haubs, and Schultz, 1986). Here, the manager is focussing above all on the potential for quality increase which cannot be exploited in conventional cultivation. It is therefore planned to first switch the best areas to organic, effective immediately, in order to get an idea of the methods involved. Here too, the high intensity of the manual labour necessary complements the personnel, quality and product concepts of the business.

The traditional and labour intensive quality approach to cultivation is also extended to the wine cultivation. The basic features of the cellars can be retained in the medium term. A return to traditional methods will provide better handling of the grapes, and it is planned to no longer transport the grapes to the winepress using pumps, but with tipping containers, and filters should as far as possible be avoided (Schneider, 2005). Ripening in wooden barrels will be accompanied by modern cooling systems, and fermentation should use wild yeast varieties to produce characteristic wines. The basic approach corresponds with the core competency concept, in that the vinery works on the assumption that quality originates in the vineyard. In the cellars, the task is to maintain this quality and allow the wines plenty of space to develop. Although the majority of new cellar technologies make the wine more sterile, they also reduce their natural character. Through a product-related quality orientation in production, the relevant measures to reach the desired end-product can be selected even in the cultivation process. It is thus possible to handle the raw materials appropriately from a product cost perspective. From this stems a wine concept consisting of several elements, planned from vineyard to customer.

Observing the new production concept, it is apparent that the working methodology is, at least in part, more traditional than it was 60 years ago. However, these classical methods are accompanied by modern analytical processes that allow early recognition of the need to intervene in the development of the wine. At this station of cultivation, the vinery tries to minimize the impact of any interference in the process. For example, they are involved in the development of a particularly protective filter. The business calls this combination of traditional approaches and modern techniques "Tradition 2.0".

3.5 Marketing and communications strategy

Correspondent to the product pyramid and the traditional production techniques, a packaging layout was chosen that reduced the historical elements to their essential minimum and then strikingly displayed these. The new labels of the premium line are thus designed in blue tones brought in back in the 1920s, and the symmetry and the code of arms are also taken from old labels of the vinery. However, through reduction of less important content, the labels nevertheless appear very modern. The labels of the basic line "ReinWeiß" take this reduction of content even further, and limit themselves to the product brand and the grape variety. Using the combination of classical fonts and modern layout, a distinctive design has been produced that is effectively conveyed through high quality printing (Hoffmann, 2009; Mello and Pires,

2009). The wines are sold in traditional Rhine wine bottles. This deliberate rejection of international bottle shapes underlines the importance of the regional and terroir links, and embodies a new self-confidence for Rhine wines. The high quality wines are sealed with natural cork, whilst a modern glass seal is intended for the middle and basic quality wines. The glass seal appears quite traditional due to its high quality production and classical material, yet is one of the most modern seals in the industry.

This approach of a sensitive combination of traditional high quality and modern style elements is carried through to the business' image in all printed media of the vinery, where a reduction in buzzwords, slogans and pictures conveys the simple concept of the product development. Special attention is paid to the design of their internet presence, and in 2009 the vinery's new webpage was launched, which in its layout and concept is considerably more professional than many in the industry (www.schaetzel.de). The site should bridge the gap between a fundamentally traditional product and a young target group, but at the same time be user-friendly enough to allow older consumers to comfortably use it. The site is furthermore a showcase for the press and thereby the basis of successful media relations.

It is important to the vinery that the restructuring, and in particular the rise in product prices, is carried out so that any customers lost are always overcompensated for by new customers gained. Thus, prices were raised gradually, whilst media work was intensified to win new customers (Castriota and Delmastro, 2008). In addition, high class events were organised, whereby the approaches of the production philosophy were once again always central and underlined by other elements. Music is, for example, mostly performed 'by hand', and thus the use of live piano music should also reflect the artisan production of the wine. In the same way, the preparation of high quality, mostly regional products on the grill provides a culinary experience with the minimal amount of technology necessary. A core element to win over new private customers is the positioning with top-quality catering.

In the past, catering has been overseen and supplied predominantly either by specialist wine dealers or agents. Here it is difficult for the as yet little known Schätzel vinery to bypass the well-established giants of the industry, so the business settled on a new approach. It entered into an exclusive contract with multi-regional wholesale drinks suppliers, which until that point had been specialised on soft drinks, beer and spirits, and for whom wine was previously only a sideline product. As, however, the wholesale drinks suppliers had well developed business contacts to the food industry, including the high-class segment, the vinery hoped for a change in the industry structure. The trend towards centralising purchases and the increasing obligations to stake partners may compel caterers and restaurateurs in the future to increasingly source their wines through these suppliers. The first wholesale drinks suppliers have already begun to employ sommelier and are investing noticeably in this area. Particularly the 'just-in-time' deliveries, potentially up to several times a week, allow those in the catering industry to reduce their capital commitment in wine stores and be more flexible and up-to-date with their wine selection.

The business manager plans to directly support the marketing through his personal involvement, and frequently attend with the sommelier. In addition, the vinery is trying with promotional material (menu card holders, wine menu inlays, information leaflets for the service) to place its products in upmarket restaurants and catering businesses. Through the central organisation of the wholesale drinks suppliers, it is possible to plan and agree long-term on the product range and batches of individual products, giving the business planning security.

4 Conclusions and outlook

On the basis of the above-described example, it becomes clear how an unprofessionally structured sideline business can be transformed into a successful functioning full-time concern. The challenge was in ensuring healthy growth whilst keeping risks to a minimum. The described procedure for the realignment of the business should not, however, be taken as a template for guaranteed success, as the Schätzel vinery also benefitted from its very advantageous existing core resources. Against the backdrop of this case study, it is also clear that next to these core resources, also the competence, creativity and innovative strength of the business manager represent key elements in the success.

The vinery's focus for the future can be defined as consistently pursuing the path that they have created. This means that the vinery must reinforce their position in the market in order to build firm foundations for the further growth to a vineyard area of 10 ha. Also at the sales level, efforts are currently being made to establish a marketing network through cooperation with other international wineries. Within this framework, the development of a new, own quality standard is being considered, which should make it possible to ensure sustainable quality standards within the cooperation.

References

Ali, H. H., Lecocq, S., and Visser, M. (2008). The impact of gurus: Parker grades and en primeur wine prices. *The Economic Journal, 118*(529), 158-173.

Castriota, S., and Delmastro, M. (2008). Individual and Collective Reputation: Lessons from the Wine Market. *AAWE Working Papers*. Retrieved January 10, 2010, from http://ssrn.com/abstract=1349992.

Gemmrich, A. R. (1998). *Nachhaltige Weinwirtschaft : Wegweiser zum betrieblichen Umweltschutz in Weinbaubetrieben, Kellereien und Winzergenossenschaften.* Neustadt a.d. Weinstraße: Meininger.

Gergaud, O., and Livat, F. (2007). How Do Consumers Use Signals to Assess Quality? *AAWE Working Papers*. Retrieved January 10, 2010, from http://www.wine-economics.org/workingpapers/.

Hanf, J., and Schweickert, E. (2007). Changes in the wine chain - Managerial challenges and threats for German wine co-ops. *AAWE Working Papers*. Retrieved January 10, 2010, from http://ageconsearch.umn.edu/handle/37315.

Hoffmann, D. (2009). *Marketing für deutschen Wein - spez. Focus: Export und Riesling.* Paper presented at the *Agrarmarketingtage* December 7-8, 2009, Berlin.

Mello, L. d., and Pires, R. (2009). *Message on the Bottle: Colours and Shapes of Wine Labels.* Paper presented at the *Fourth International Conference of the Academy of Wine Business Research*, Siena.

Payne, J. (2010). *Gault Millau WeinGuide Deutschland 2010: Der Weinführer für Genießer.* München: Christian Verlag.

Scheuermann, M. (2009). Winzer-Markenweine im Test. *Vinum.*

Schneider, V. (2005). Postfermentative Phase: Die Hefe nach der Gärung. *Der Winzer, 11.*

Wörner, M. (1984). *Untersuchungen über arbeitextensivere Formen der Rebenerziehung in Steillagen unter besonderer Berücksichtigung der Umkehrerziehung.* Unpublished report. Geisenheim.

Wucherpfennig, K., Haubs, H., and Schultz, H. R. (1986). Qualitätsbewußter und ökologisch orientierter Weinbau *KTBL-Schrift.*

Part III

New avenues of research: online marketing and sensory marketing

Institutional isomorphism and adoption of e-marketing in the hospitality industry: a new perspective for research

Amos Gyau
Randy Stringer

School of agriculture, food and wine
Discipline of Agri-food and Wine Business
The University of Adelaide, Australia
{Amos.gyau, randy.stringer}@adelaide.edu.au

Abstract

Many factors have been used to explain adoption behaviours of managers in the hospitality industry. Studies employing these factors have been carried out with the assumption (implicit or explicit) that the managers are free and independent in their decision to adopt innovations such as the use of e-marketing without much attention to the effects of external pressures. In this paper we provide a relatively new conceptual framework for innovation adoption in the field of tourism based on the concept of institutional isomorphism. We argue that e-marketing adoption decisions by tourist operators can be influenced either singly or jointly by important aspects of institutional factors namely 'institutional isomorphism', which, when combined with the traditional innovation adoption factors can enhance a better understanding of tour operators adoption behaviour.

Zusammenfassung

In der Literatur finden sich viele Faktoren, die das Innovationsverhalten von Managern im Gastgewerbe erklären. Studien, die diese Faktoren nutzen, sind mit der (impliziten oder expliziten) Annahme durchgeführt worden, dass Manager ihre Entscheidungen, Innovationen wie z. B. E-Marketing einzuführen, frei und unabhängig treffen können, ohne äußeren Druck berücksichtigen zu müssen. In diesem Beitrag bieten wir ein relativ neues konzeptionelles System für die Einführung von Innovationen in der Tourismus-Branche, das auf dem Konzept der institutionellen Isomorphie beruht. Es wird argumentiert, dass die Entscheidungen von Reiseveranstaltern bzgl. der Einführung von E-Marketing entweder durch einzelne oder mehrere institutionelle Faktoren beeinflusst werden können. Dem Konzept der institutionellen Isomorphie kommt hierbei eine besondere Bedeutung zu. Kombiniert mit traditionellen Faktoren der Innovations-Übernahme kann sie das Verständnis des Innovationsverhalten der Reiseveranstalter verbessern.

Riassunto

Sono stati usati molti fattori per spiegare i comportamenti adottati dai manager nell'industria dell'ospitalità. Si son fatti degli studi che utilizzano tali fattori con il presupposto - implicito o esplicito- che i manager siano liberi e indipendenti circa la decisione di adottare delle innovazioni quali ad esempio l'utilizzo dell'e-marketing. Poca attenzione è stata data a eventuali pressioni esterne e alle loro conseguenze. In questo lavoro forniamo un quadro concettuale relativamente nuovo per l'adozione di innovazioni nel campo del turismo basato sul

K.L. Sidali et al. (eds.), *Food, Agri-Culture and Tourism: Linking local gastronomy and rural tourism: interdisciplinary perspectives*, DOI 10.1007/978-3-642-11361-1_9,

concetto dell'isomorfismo istituzionale. Crediamo che le decisioni di adottare l'e-marketing da parte degli operatori turistici possano essere influenzate singolarmente o collettivamente da un importante complesso di fattori istituzionali detti "isomorfismo istituzionale". Quest'ultimo, insieme ai tradizionali fattori che spiegano il comportamento relativo alle innovazioni, può aiutare a meglio comprendere perché i tour operators adottino tali comportamenti.

Keywords: Institutional isomorphism; Tourism; e-marketing adoption.

1 Introduction

The question of whether innovation in tour operations is important will undoubtly receive a positive response from academics, tour operators, visitors and policy makers alike. The use of internet marketing has appeared over the past decade in many industries including the banking sector, hospitality as well as retailing and distribution. The e-marketing permits strong goal orientation, personalisation, individualisation and the most efficient measurability of advertisement campaign's effectiveness (Barowski, 2000). Furthermore, a number of studies have identified a positive and significant relationship between the use of information technology (IT) and competitive advantage (e.g., Cho and Olsen, 1998; Porter, 1985). Consequently, many businesses have recognised the use of the e-marketing as a very important area of IT innovation and investment (Sauer, 2000). Given the potential benefits of the IT as discussed in the literature, the extensive use of IT in the hospitality industry would appear to be a foregone conclusion (Siguaw and Enz, 2000). However, as further noted by Siguaw and Enz (2000), the degree to which the hospitality industry has embraced technological innovations have not been consensual leading to incomplete knowledge and understanding of the adoption behaviours of managers in the tourism industry.

Lai et al. (2006) have argued that most adoption studies have implicitly assumed that innovation adoption is purely an independent phenomenon which actors are free to choose whether to adopt or not. However, the authors noted that people's (tourist operators in this case) freedom of choice of a technology may be limited by certain institutional isomorphic pressures which may influence the operators to adopt certain types of innovations even when there is no direct economic justification for doing so. Against this background, the concept of institutional isomorphism which causes one organisation to resemble others in an industry (Dimaggio and Powell, 1983) may be considered as an important candidate that can be used to further highlight tour operators' decision to adopt e-marketing.

The aim of this paper is to complement the existing literature on innovation adoption of e-marketing especially in the field of tourism by introducing the concept of isomorphism as an important aspect of institutions that can be combined with the traditional adoption factors to explain tour operators' adoption behaviour. The paper systematically discusses and relates three main processes of institutional isomorphism to the adoption of e-marketing in the tour industry. The possible impacts of the various forms of institutional isomorphic pressures discussed can enable stakeholders to better understand how the institutional pressures they exert on operators affect their adoption behaviours.

The remaining sections of the article are organised as follows: in the next section, we provide a brief review of the literature on innovation adoption in tourism. We then discuss the concept of institutional isomorphism and show how the isomorphic processes can influence the level of adoption within the field of tourism. Finally, a conceptual framework is developed, conclusions are drawn and direction for future research provided.

2 Defining e-marketing

E-marketing refers to the use of internet technology to conduct business (Lin and Lin, 2007). Different terms, such as e-business, e-marketing or e-commerce have been used in the literature and as such will be considered synonymous for the sake of this study. E-business is different from the previous traditional technological innovations such as the management information systems and the electronic data interchange which were used to improve the efficiency of the IS system and supports batch exchange of structured procurement documents (Moore and Benbasat, 1991).

As a means of commercial communication, e-marketing uses email to deepen the relationship with current or past customers in order to increase their loyalty (Bourland and Handley, 2000). According to Zhu (2004), e-business represents a new way to integrate internet based technology with core business, potentially affecting the whole business. The most frequently used types of e-marketing are newsletters, commercial emails and viral marketing (Fuchs et al., 2008)

3 Previous research on adoption of e-marketing

Technology adoption presents a complex process which has been studied widely and intensively from different perspectives (Ghadim and Pannell, 1999; Novoselova et al., 2007). Subsequently, many studies have reviewed comprehensively the literature on innovation adoption in general (e.g., Rogers, 2003) and for e-business in particular (e.g., Fuchs et al., 2008; Lin and Lin, 2007).

Technology Organisation Environment (TOE) model is one of the most prominent theories that have been used to study the e-marketing adoption in prominent adoption models. The TOE model is based on the work of Tornatzky and Fleischer (1990). The authors identified factors such as the technological context, organisational context and the environmental context as determining the adoption process. The technological context defines the internal and external environment relevant to the firm, the organisational context considers the firm size, centralisation and formalisation, complexity of leadership structures as well as the resources that are available to the firm. The environmental factors is related to the economic as well as the competitive structures.

Davis (1989) developed the Technology Acceptance Model (TAM) which focuses on decision makers attitudes and beliefs. Davis et al. (1989) argued that adoption decision is significantly influenced by individual's intention to use the technology.

Fuchs et al. (2008) studied the factors affecting the adoption behaviour of e-marketing in the alpine hotels and caibrated three models involving previous adoption, present adoption and intention to adopt e-marketing in the future. The authors observed that managers' adoption decision is influenced by factors such as environmental conditions, organisational, perceived techical competence, perceived pressure from customers and competitors, conditions, innovation conditions as well as the individual cosmopolitarianism.

Siguaw and Enz (2000) examined the utilization of information technology in all hotel sectors (from deluxe to budget) in the US using two unique independent samples. Their studies revealed that overall, the US lodging industry has adopted technologies that improve the performance of employee productivity and enhance revenue but has not given priority to technologies designed to improve guest service. Further, the authors observed that the hotel sector, lodging types, complexity of the property and independent versus chain operations influence the number and type of technology adoption.

In summary, it can be observed that various factors have been used to explain adoption behaviour in general and in e-marketing in particular. However, these studies have not explicitly discussed and integrated the idea of institutional isomorphic processes into their models. In situations where the underlying economic motives of these sociological terms have been considered in the e-marketing literature, it has been selective and fragmented and therefore models that have considered aspects of isomorphism have not been integrated and comprehensive.

Our approach takes a new perspective to complement the existing literature with a discussion of the concept of isomorphism and how it can influence e-marketing adoption decisions in the tourism sector. In the subsequent sections, we provide a general overview about the concept of the institutional isomorphism and discuss the three main institutional isomorphic processes. Following this, we explain how institutional isomorphic processes can impact on e-marketing adoption.

4 Institutional isomorphism

4.1 Defining institutional isomorphism

This article draws on the concept of isomorphism to explain adoption behaviour of tour operators and managers in the hospitality industry. The proponents of institutional isomorphism concept argue that the process of isomorphism and/or legitimacy is the central force explaining why and how several practices (such as technology adoption) have emerged and changed (Hassan, 2008). The institutional theory recognises the influence on organisations and pressures which compel organisations to behave in certain defined ways. The proponents of the theory have identified a number of different institutional elements and institutional view of isomorphism proposed by DiMaggio and Powell (1983) is one of them (see also Scott, 1987; Scott and Meyer, 1992). According to DiMaggio and Powell (1983), it is not only economic reasons that drive the development of new systems, but rather the seeking of social legitimacy. Accordingly, the proponents distinguish between two main forms of isomorphism, namely; competitive isomorphism and institutional isomorphism (DiMaggio and Powell, 1983). Competitive isomorphism is concerned with efficiency, whereby the adoption of certain practices or technology such as e-marketing is mainly based on its economic justification (i.e., provides better economic benefits) (Greenwood and Hinnings, 1996). Institutional isomorphism on the other hand is concerned with the social forces that put pressures on organisations (such as tour operators in this case) to adopt certain practices (DiMaggio and Powell, 1983). The theory attempts to address the circumstances that cause actors who recognise and try to act on their interests to be unable to do so effectively due to pressures from other organisations. The basic premise of the DiMaggio and Powell (1983) theory is that organizational fields follow an evolutionary path from diversity to homogeneity, which is conceptualised as isomorphism. In line with Hawley (1968), they described isomorphism as a constraining process that forces one unit in a population to resemble other units that face the same set of environmental conditions. New institutionalism recognizes that institutions operate in an environment consisting of other institutions, called the institutional environment. Every institution is influenced by the broader environment (or in simpler terms institutional peer pressure). In this environment, the main goal of organizations is to survive. In order to do so, they need to do more than succeed economically by establishing legitimacy within a given field. Thus, it gets to a point in the development of an organization where the primary objective of an organizational change is not to better economic performance (at least not in the short term) but to obtain greater legitimacy. In this context, it is argued that organisations can adjust their internal characteristics in order to conform to the expectations of the key stakeholders in their environment (Ashworth et al., 2007).

Institutional theory has become an important theoretical lens within which processes of change in organisations are explained and interpreted. Lai et al. (2006) for instance adopted the institutional isomorphism perspective to explain the adoption of information technology for supply chain management and explained the possible impacts of the three main institutional isomorphic pressures. Mir and Rahaman (2005) adopted the concept of institutional isomorphism to explain the adoption of International Accounting Standards (IASs) in Bangladesh. These authors observed that institutional legitimisation is a major factor that drives the decision to adopt because of the pressure exerted by key international donor/lending institutions on the Bangladeshi Government and professional accounting bodies. According to them, such pressure results from not only the need to provide credibility to foreign investors but also the need for strong accountability arrangements with lending/donor agencies.

Institutional theorists emphasize the significance of social and cultural aspects in organisational environment. Organizational environment are characterised as the sources of norms and values that permeate organisations and influence actions and assumptions regarding the behaviours, organisational forms and processes that are seen as legitimate (Ashworth et al., 2007).

4.2 Processes of institutional isomorphism

DiMaggio and Powell (1983) identified three main ways in which institutional pressures occur. These are described as mimetic forces, coercive forces and normative forces. The mimetic forces are the pressures to copy other organisations' activities, systems, or structures. This aspect of isomorphism assumes that when firms have ambiguous goals and operate in a volatile environment, they model themselves on other organisations, especially in closely related fields or industries in response to the uncertain business environment (Lai et al., 2006). Mimetic isomorphism occurs as a field emerges in its formative phase or during a reformulation phase brought about by the introduction of a major innovation, which leads to a high level of uncertainty. In response, actors survey the environment and borrow legitimised practices from others, especially from the superior performing actors in the field. These practices include benchmarking and identification of best practices and leading players in the field.

Coercive institutional isomorphism occurs when pressure is exerted on a dependent firm by other organisations and by cultural expectations in the society within which the dependent firm operates (DiMaggio and Powell, 1983). These forces are usually exerted by governments, regulatory or other agencies to adopt the systems that they favour and are normally related to legal requirements, health and safety regulations but may also stem from contractual obligations with other actors which constraint organisational variety.

The third form of institutional isomorphic pressures identified by DiMaggio and Powell (1983) is the normative isomorphism. These pressures describe the effects of professional standards and the influence of professional communities on organisational characteristics. They capture the ways in which organisations are expected to conform to standards of professionalism and to adopt systems and techniques considered to be legitimate by relevant professional groupings. According to Lai et al. (2006), these norms are conveyed through education and training of professionals and certification process accredited by professional bodies.

5 Discussion

In this section, we discuss how isomorphism processes can influence e-marketing adoption in the hotel and tourism industry.

Coercive isomorphism for instance can be considered as one of the most powerful pressures that may influence operators to adopt e-marketing. This is particularly true against the backdrop of increasing tourist concern and demand for better services which are fuelling the emergence of standards in tourism operations. These concerns, which may be real or perceived, may generate a diversity of responses from operators to adopt various forms of standards and innovations, such as the internet based marketing activities and promotions, which may be perceived by tourists to be more effective and convenient. These standards are often regulated and enforced by specialised bodies and organisations.

The adoption of the e-marketing can also be explained by the existence of normative pressures, which may result from customers and competitors (Zhu and Kraemer, 2002). As many other tour operators adopt e-marketing in their activities, other operators are also likely to follow suit as the e-marketing technology might be perceived by the industry as the acceptable way of doing business. Consequently, the technology may become entrenched to the extent that any operator who might decide to do otherwise becomes anomalous in the tourism industry. In an attempt to obtain legitimacy among their peers as well as regulators and other stakeholders, managers in the hotel and tourism sector may decide to join such a scheme even when doing so does not provide direct economic and business benefits. In this context, the so called "bandwagon" and "neighbourhood" effects as discussed by Michihiro and Rob (1998) and Langyintuo and Mekuria (2008) respectively may influence managers to adopt thereby creating semblance among the operators in terms of what they adopt. Another source of such normative forces may come from training courses which usually provide operating norms, technology and standards to the managers. These activities usually provide similar forms of information to the operators and therefore provide incentives for the participants to adopt similar innovations.

The final form of isomorphic process that may be employed to explain why managers may adopt innovations such as e-marketing is the mimetic isomorphism. This particularly happens in situations where some managers may copy other managers who they perceive to be more successful in their operations. This happens when there is a high level of uncertainty about a particular innovation such as the e-marketing. Thus, new operators for instance might want to model their business operations along the lines of the old and existing businesses in the tourism industry which are also adopting the e-marketing. This particularly happens in situations of ambiguity of decision process in what Abrahamson and Rosenkopf (1993, p. 494) referred to as "opaqueness or lack of clarity surrounding organisational assessment of an innovation". According to them, when the adoption process is uncertain and results are ambiguous, there might be what they called "bandwagon competitive pressures" toward the introduction of an innovation. Bandwagon occurs if potential adopters perceive the risk that innovation is a success, and that if they do not adopt it, their performance will fall below the average performance of adopters. March and Olsen (1976) also support how mimics can induce adoption by arguing that as companies cannot evaluate the technical efficiency or the results of an innovation, because of the ambiguity of goals, means-ends relations and environment, it will rely more on social as opposed to economic factors in their decision on whether or not to adopt.

The above discussions are further synthesised in Figure 1. The model shows the processes that lead to adoption of technology and the impact of such an adoption.

Figure 1. Conceptual model for institutional is omorphism and e-marketing adoption

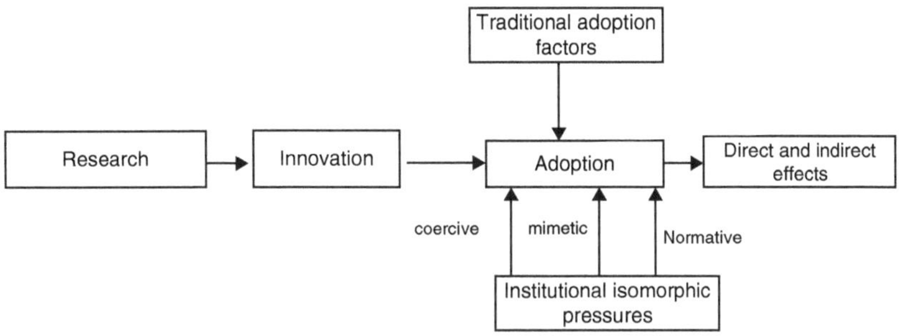

Source: Own development

It indicates that both the traditional adoption factors as discussed in our literature review on the one hand and the institutional isomorphic pressures on the other hand can singly or jointly influence the level of adoption.

6 Conclusion and future research

In this paper, we have explored how the three main isomorphic processes, namely coercive, mimetic and normative pressures, can influence hotel and tourism business managers' free will and independence in their decision to adopt e-marketing. The article contributes to the technology adoption literature in the tourism industry by extending our knowledge and elaborating on aspects of institutional factors that influence technology adoption namely the 'Institutional Isomorphism processes'. The model presented does not intend to provide an alternative way of explaining managers' adoption behaviour but rather complements and further elaborates on aspects of institutional factors which have not been addressed adequately in the tourism literature. The model therefore provides a basis and additional impetus for future researchers in the field of technology adoption in the tourism industry to further develop and provide empirical evidence on the relative importance of the institutional isomorphic processes and the traditional adoption factors as discussed. In this way, we would be able to have a holistic understanding and a more elaborate way of explaining managers' adoption behaviour. In this context, questions which may be of interest for future research may include: What are the relative importances of isomorphic pressures and the traditional adoption factors in explaining tourism related innovations such as the e-marketing? How can we distinguish between the two? What are the main sources of institutional isomorphism which affect tour operators' adoption behaviour? What is the relative strength of each of these processes? What is the main motivation for operators to adopt e-marketing? The above questions will provide an important platform for future researchers to be able to further understand the motivation and determinants of e-marketing adoption in the field of tourism.

References

Abrahamson, E., and Rosenkopf, F. (1993). Institutional and competitive bandwagons: using mathematical modeling as a tool to explore innovation diffusion. *Academy of Management Review, 18*(3), 289-309.

Ashworth, R., Boyne, G., and Delbridge, R. (2007). Escape from the iron cage? Organisational change and isomorphic pressures in the public sector. *Journal of Public administration Research and Theory*. Retrieved December 21, 2007, from http://jpart.oxfordjournals.org/cgi/content/abstract/mum038.

Barowski, M. (2000). *Online-marketing* (1st ed.). Berlin: Cornelsen.

Bourland, A., and Handley, A. (2000). E-Mail Marketing Secrets. Retrieved September 10, 2009, from www.emigrantas.com/books/marketing/(business%20ebook)%20%20 Email %20Marketin^%20Secrets%2Oli.pdf.

Cho, W. and Olsen, M.D. (1998). A case study approach to understanding the impact of information technology on competitive advantage in the lodging industry. *Journal of Hospitality and Tourism Research, 22*(4), 376-394.

Davis, F.D., Bagozzi, R.P. and Warshaw, P.R. (1989). User acceptance of computer technology. A comparison of two theoretical models. *Management Science, 35*(8), 982-1003.

Davis, F.D. (1989). Perceived usefulness, ease of use and user acceptance of information technology. *Management Information Systems Quarterly, 13*(3), 318-340.

DiMaggio, P.J. and Powell, W. (1983). The iron cage revisited: institutional isomorphism and collective rationality in organisational fields. *American Sociological Review, 48*, 147-160.

Fuchs, M., Tuta, M., and Höpken, W. (2008). Adoption of Email marketing in the hotel sector. In *Information and Communication Technologies in Tourism 2008* Proceedings of the International Conference in Innsbruck, Austria, 2008. Wien: Springer.

Ghadim, A.K., and Pannell, D.J. (1999). A conceptual framework for the adoption of an agricultural innovation. *Agricultural Economics, 21*(2), 145-154.

Greenwwod, R., and Hinnings, C.R. (1996). Understanding radical organisational change: Bringing together the old and new institutionalism. *Academy of Management Review, 21*(4), 1022-1054.

Hawley, A. (1968). Human ecology. In David L. Sills (Ed.), *International Encyclopaedia of the Social Sciences* (pp. 328-337). New York: MacMillan.

Lai, K.-H., Wong, C., and Cheng, T.C. (2006). Institutional isomorphism and the adoption of information technology for supply chain management.*Computers in Industry, 57*, 93-98.

Langyintuo, A., and Mekuria, M. (2008). Assessing the influence of neighbourhood effects on the adoption of improved agricultural technologies in developing agriculture. *African Journal of Agricultural and Resource Economics, 2*, 151-169.

Lin, H.-F., and Lin S.-M. (2007). Determinants of e-business diffusion: A test of the technology diffusion perspective. *Technovation, 28*, 135-145.

March, J.G., and Olsen, J.P. (1976). *Ambiguity and choice in organisations*. Bergen: Univertetsforlaget.

Michihiro, K., and Rob, F. (1998). Bandwagon effects and long run technology choice. *Games and Economic Behaviour, 22*, 30-60.

Mir, M.Z., and Rahaman A.S. (2005). The adoption of international accounting standards in Bangladesh: An exploration of rationale and process. *Accounting, Auditing and Accountability Journal, 18*(6), 816-841.

Moore, G.C., and Benbasat, I. (1991). Development of an instrument to measure the perceptions of adopting an information technology innovation. *Information Systems Research, 2*(3), 192-222.

Porter, M.E. (1985). *Competitive advantage: Creating and sustaining superior performance*. New York: Free Press.

Rogers, E.M. (2003). *Diffusion of innovations* (5th ed.). New York: Free Press.

Sauer, C. (2000). Managing the infrastructure. In Willcocks, L., and Sauer, C. (Eds.), *Moving to e-business.* London: Random house.
Scott, W., and Meyer, J. (1992). The organisations of social sectors. In Meyer, J., and Scott, R. (Eds.), *Organisaional environment: Ritual and rationality* (pp. 129-153). Beverly Hills: Sage.
Scott, W. (1987). The adolescence of institutional theory. *Administrative Science Quarterly, 32,* 493-511.
Siguaw, J.A., and Enz, C.A. (1999). Best practices in information technology. *Hotel and Restaurant Administration Quarterly, 40*(10), 58-71.
Tornatzky, L., and Fleischer, M. (1990). *Process of technology innovation.* Lexington: MA.
Zhu, K., and Kraemer, K.L. (2002). E-commerce metrics for net enhanced organisations: Assessing the value of e-commerce to firms' performance in the manufacturing sector. *Information Systems Research, 13*(3), 275-295.
Zhu, K. (2004). The complimentarity of information technology infrastructure and e-commerce capability: a resource based assessment of their business value. *Journal of Management Information Systems, 21*(91), 167-202.

E-Business horizons in the tourism industry - Challenges for research and practice

Matthias Fuchs[a/b]
Wolfram Höpken[a/c]

[a] eTourism Competence Center Austria (ECCA)
University of Innsbruck, Austria

[b] European Tourism Research Institute (ETOUR)
Mid-Sweden University, Sweden
matthias.fuchs@miun.se

[c] Business Informatics Group
University of Applied Sciences Ravensburg-Weingarten, Germany
wolfram.hoepken@etourism-austria.at

Abstract

The discussion paper highlights challenges for the travel and tourism industry provoked by new information and communication technologies (ICTS) as well as related veins of e-Business research. First of all, the paper explores stylized facts from the last 15 years of web history in tourism and delineates actual levels of ICT adoption in tourism. Moreover, behavioural and structural changes on account of ICT in tourism are explained by adopting a multi-disciplinary approach. Finally, an innovative ICT application from tourism practice – the destination guide www.dolomitiSuperski.mobi is portrayed. In the outlook the paper sketches the future of e-Tourism research by highlighting the interrelation between computer, information and management sciences.

Zusammenfassung

Hervorgerufen durch neue Informations- und Kommunikationstechnologien (IKT) beleuchtet das folgende Diskussionspapier neue Herausforderungen für die Reise- und Tourismusindustrie sowie bedeutende Bereiche der e-Business Forschung im Tourismus. Der Beitrag ist wie folgt aufgebaut: Abschnitt zwei gibt einen Überblick über wichtige Fakten zur Erfolgsgeschichte des Internet-Tourismus während der letzten 15 Jahre. Darüber hinaus werden Trends der IKT Nutzung im Tourismus aufgezeigt. In Abschnitt drei werden verhaltensspezifische und strukturelle Veränderungen sowie Anpassungsprozesse aufgrund von IKT Nutzung im Tourismus durch Heranziehung eines multidisziplinären Ansatzes erklärt. Abschnitt vier stellt eine innovative IKT-Anwendung aus der Tourismuspraxis vor – das mobile Informationssystem www.dolomitiSuperski.mobi. Unter besonderer Berücksichtigung der Wechselbeziehung zwischen Computer-, Informations- und Managementwissenschaften skizziert der letzte Abschnitt wesentliche Forschungstrends des E-Tourismus.

K.L. Sidali et al. (eds.), *Food, Agri-Culture and Tourism: Linking local gastronomy and rural tourism: interdisciplinary perspectives*, DOI 10.1007/978-3-642-11361-1_10,

Riassunto

Il presente lavoro esamina le sfide per l'industria dei viaggi e del turismo poste dalle nuove tecnologie dell'informazione e della comunicazione (ICTS) come pure dai risultati ottenuti dalla ricerca e-business. La relazione è così strutturata: la seconda sezione fornisce un rapporto esplorativo sui fatti più rimarchevoli riscontrati negli ultimi 15 anni di cronaca informatica nel turismo. Inoltre, sono delineati gli attuali livelli di ICTS adottati per il turismo. Nella terza sezione sono spiegati i cambiamenti comportamentali e strutturali riguardanti l'uso dell'ICTS nel turismo, mediante un approccio pluridisciplinare. La quarta sezione mostra una recentissima applicazione ICTS nella realtà turistica: il servizio mobile www dolomiti.mobi. La sezione relativa alle prospettive future dà un primo quadro della futura ricerca nel e-turismo, chiarendo le rispettive connessioni tra computer, informazione e scienze manageriali.

Keywords: e-Business; ICT adoption, Mobile guide; e-tourism research.

1 Introduction

During the last 15 years information and communication technologies (ICT), and in particular the appearance of the Internet, have stimulated radical changes in the operation of tourism businesses as well as entire destination value chains (Werthner and Ricci, 2004; Buhalis and O'Connor, 2005; Gretzel and Fesenmaier, 2009). The speed of Internet development can be described impressively by the quintupling of global users during the last decade. While in 2000, worldwide 400 million individuals made regular use of the Internet, the corresponding figure is estimated at two billion in 2011 (Werthner, 2009). Today, the Internet contains 130 million distinct web-sites, 40 billion web pages and one trillion links. Moreover, the information within the Internet is quantified at 160 Exa (i.e., 10^{18}) bytes (ibid., 2009). However, the majority of the existing information is replicated, with only 25% representing original content. From an application point of view, the need for effective and innovative query mechanisms emerges from the fact that 95% of Internet data is unstructured and loosely distributed over the globe (Fesenmaier et al., 2004). Finally, from a business perspective, the Internet is denoted as the main driver behind value creation in service networks since it facilitates user integration and makes complex transactions and interactions more effective (Porter, 2001; Zhu et al., 2005).

The aim of this discussion paper is to highlight major challenges for the travel and tourism industry provoked by new ICTS as well as related veins of e-Business research. The paper is structured as follows: Section two provides an explorative report on stylized facts from the last 15 years of web history in tourism. Moreover, actual levels of ICT adoption and diffusion in travel and tourism are delineated. In section three behavioural and structural changes on account of ICT in tourism are discussed by adopting a multi-disciplinary framework. Section four portrays a case-study of an innovative ICT application from tourism practice – the destination guide www.dolomitiSuperski.mobi. The outlook section sketches the future of e-Tourism research.

2 Stylized facts of ICT in travel and tourism

Apart from the current economic downturn, travel and tourism industries typically show an average increase well above the global level of economic growth (UNWTO, 2009). Moreover, travel services and holiday products constitute one of the most expensive items purchased regularly by households, thus representing a significant portion of individual's annual budget (Buhalis and Law, 2008; p. 611). By the same token, the ICT industries are worldwide known as the fastest growing branches of economic activity, comprising such sub-branches like technical consulting, Internet publishing and broadcasting, wireless telecommunication, computer systems design, data hosting and processing (Forbes, 2008). Thus, not surprisingly, within the e-Commerce sector travel and tourism represents the largest single business sector. In 2009, for instance, 25.7% (i.e., 65.2 Bn. €) of the European total online sales volume has been generated by the travel and tourism sector, whereat in 2001 this figure stood at 5 Bn. € (Marcussen, 2009). It is noteworthy that 55.3% of actual tourism online sales are generated by the airline industries, whereas 18.6% and 14.6% may be directed to the hotel and tour operator sector, respectively. Finally, 7.4% and 4.1% pertain to the train and to the hired car sector (ibid., 2009). A comprehensive picture about ICT adoption for electronic business in the European travel and tourism industry is provided by eBusinessW@tch (2006). Electronic business (e-Business) is defined as 'the totality of an organization's computer supported external and internal business processes' (ibid., 2006; p. 191). Accordingly, tourism shows a relatively high rate of ICT penetration with respect to applications for web-based marketing and online distribution. For instance, 98.2% of Austrian hotel businesses offer a website, and 68% of them also feature booking functionalities (Fuchs et al., 2009). Moreover, 96% of Austria's hotel businesses distribute their accommodation packages externally through online platforms. Of further interest is the observation that hotel companies generate on average 17.6% of their sales through online platforms and 13.2% through owned bookable websites (ibid., 2009). Clearly, this trend is underscored also by demand side figures: before booking a travel product, 95% of Internet users conduct their information search online thereby visiting on average 10 different web sites (Farrokh, 2009). Moreover, 54% of the Internet users book leisure trips and tourism services online (Pyka and Freitag, 2009). However, shortcomings in travel and tourism become evident with respect to online business networks (e.g., supply-chains) and integrated process automation (eBusinessW@tch, 2006). Again, this notion can be confirmed by the example of the Austrian hotel sector, as only 40% of hotel businesses adopted yield management, enterprise resource planning or a personal information system (Fuchs et al., 2009).

A further observation is devoted to the market structure (i.e., concentration level) of electronic tourism markets. Interestingly enough, only three multi-national companies, namely, EXPEDIA^{INC} (e.g., www.expedia.com, www.tripadvisor.com and www.hoteldiscount.com), the Global Distribution System TRAVELPORT (e.g., www.galileo.com, www.ebookers.com, www.orbitz.com and www.octopustravel.com) and the Computer Reservation System Sabre-Holdings® (e.g., www.travelocity.com, www.lastminute.com, www.med.hotels.de) dominate 93% of the U.S.

electronic tourism market (Fuchs and Höpken, 2008; p. 249). Since the top 1% of worldwide existing travel and tourism websites provide for almost 55% of total users, this supply side finding may be paraphrased by the catchword 'the winner takes it all' (Werthner, 2007).

In the course of the last decade, travel and tourism underwent a dramatic change of strategic goals associated with ICT (Fuchs and Höpken, 2008, p. 253). While from 1996 to 1999 the strategic objective was to 'establish online presence' (e.g., through the launch of web sites and co-operations with booking portals), the strategic goal from 1999 to 2003 was to 'acquire new customers' by offering personalized information and web-services. From 2003 to 2006 the strategic goal became the 'retention of the customer'. The latter goal was achieved by target-specific customer response and electronic customer care. Since 2006, however, the main strategic goal is 'customer integration' and, as such, it can be strikingly described as the strategic shift 'from customer-focused to customer-driven'. This new strategic orientation is largely affected by the enormous opportunities provided by Web 2.0 technologies. Accordingly, the separation between web user and editor is increasingly ceasing to exist (Conrady, 2007; Xiang and Gretzel, 2009).

Put differently, through the provision of freely accessible authoring tools formerly static websites become predominantly dynamic. Thus, valuable 'content' about consumption experiences to be shared with others is gradually 'generated by the user' through product evaluations (i.e., ratings and reviews; Gretzel and Yoo, 2008; Sidali et al., 2009), recommendations (Jannach et al., 2009) as well as blogs and podcasts. The resulting knowledge-base, characteristically called as '*hyper-linked collective intelligence*', is heavily utilized by communities interested in specialized (e.g., tourism and leisure related) interaction qualities. Thus, through permanent and collaborative contributions from various community members the Internet has become a 'social network', having created a new type of empowered users. This new consumer is in control and has become a crucial player in the process of creating and shaping brands as well as its own experiences. Thus, tourism brands are no longer seen as identifiers but become themselves sources of experiences by evoking sensory, affective, creative and lifestyle-related associations (Gretzel and Fesenmaier, 2009, p. 571).

According to Compete, Inc. (2006), about 50% of online travel purchasers visited an online community website for their online travel purchasing and 25% posted a review on a review site after their purchase. Today, TripAdvisor represents the world largest and most successful social networking and community site in tourism comprising over 25 Mio unique users (O'Connor, 2008). The platform facilitates the reviewing of hotels around the world and brings together individuals in discussion forums and provides users with independent travel reviews and comments. Table 1 provides an overview of prominent community sites in travel and tourism (Dippelreiter et al., 2007).

Table 1. Prominent community web sites in travel and tourism

Community Site	Visits (April 08)	Links	Shared Goal	Specific features/web 2.0 technologies
www.lonelyplanet.com	1.158.800	630	Travel info	Forum, blogs for professional authors, mobile services (POI & Reviews)
www.travelpod.com	657.000	991	Travel experiences	Blogs (i.e. travelogues) associated to geographical maps (linked in mashups), mobile blogging & podcasting
www.virtualtourist.com	1.154.400	3.277	Unbiased fellow-based travel guides	Descriptions & rateable reviews of travel locations, several forums, 'post a question', reservation and booking functionalities (travel deals)
www.travelistic.com	90.000	170	Travel info & video	Flash player-based, user and professional video content is linked to geographical info & map, can be rated, commented and tagged (á la YouTube)
www.wikitravel.org	723.400	2.036	Free travel guide	Open create & edit functions, Blogs, RSS-feeds, tag clouds
www.travel.yahoo.com/trip	n.a.	n.a.	Create & illustrate custom trip plan	Schedule (timeline), journal (reviews, descriptions, podcast) & maps (integrates)
www.couchsurfing.com	900.400	539	Network for private accommodation	Messenger, contact list, blogs & reports, profiles
www.tripadvisor.com	4.413.200	7.070	Unbiased fellow reviews in forums (> 10 Mio. reviews generate 17 Mio. trips / week)	Popularity index, traveller articles (wiki-based), game travel IQ (find locations on map), right now (community events in real time), forum, podcast
www.travelblog.org	575.800	1.101	Travel info & experiences	Geographically associated blogs, forum and photos (tips, facts, history), map and routes

Referring to travel and tourism, a further example for Web 2.0 services is given by so called '*mushups*'. They support the individual composition of personal and geographical data, thus, allowing geo-referenced enquiries and content generation (Fuchs and Höpken, 2008; p. 254; Zanker et al., 2009). Empowered by geographical information systems (GIS) and global positioning system technology (GPS) in combination with information available on the Internet, these advanced mobile applications create value through location and time-specific information. Understandably, the *multiple-accessibility* to the various Web 2.0 services becomes an outstanding criterion for the acceptance of mobile technologies in travel and tourism. Since *mobile information services* will dramatically change tourists' decision and consumption behaviour (Rasinger et al., 2007), the newly developed and recently launched mobile service www.dolomitiSuperski.mobi is presented as a tourism case study in section four of this paper.

3 The multi-disciplinary nature of e-tourism research

The benefit tourism gains from ICT can be traced back to the service characteristics of the tourism product itself. Since tourists have to schedule and to configure their individualized (but not a priori testable) tourism service bundle before it can be experienced, highest interaction and information quality is demanded by consumers from the various suppliers along the entire tourism value chain (Fuchs and Höpken, 2008; p. 249). Consequently, tourism firms strive to visualize their service concepts for specific target segments through web-based multi-media applications. Moreover, electronic data interchange facilitates distribution and transaction processes (i.e., online reservation, booking, ticketing and payment, respectively). Finally, coordination of the many tourism supply elements (e.g., transportation, accommodation, sports, etc.) jointly offered by (usually small-sized) tourism firms is supported by electronic networks since they effectively facilitate collaboration, negotiation and product bundling. As a consequence, tourism service processes are demanding large amounts of data as well as huge information and communication processing efforts. Clearly, this *information intensiveness* of the travel and tourism sector was driven by the emergence of the Internet. Technically speaking, websites are neutralizing the trade-off between the number of achievable persons at the one hand, and the level of information richness (e.g., information accuracy, customization level, timeliness) at the other hand. Moreover, the Internet combines and integrates as equivalent factors information, communication, interaction, collaboration and transaction processes (Gretzel and Fesenmaier, 2009). Thus, traditional media (as well as related communication channels) will not be able to generate similar network effects, wherefore they are becoming less important. Before going on to present a framework for e-Tourism research, prominent research topics are briefly discussed. According to Buhalis and Law (2008) thematic trends in e-Tourism research are briefly described along three dimensions: *consumers and demand, technological innovation* and *industry functions*.

The topical dimension *consumers and demand* postulates that ICT places the consumer in the middle of the service delivery process (ibid., 2008; p. 610). Moreover, travellers have access to a large global pool of information provided by tourism organisations, private enterprises and - increasingly - by other consumers. However, such systems provide value for the consumer only if they are also capable of reducing the information overload typically arising from too many choices. Thus, a first vein of e-Tourism research focuses on the design, implementation and testing of pro-active services with the ability to intelligently offer (i.e., recommend) personalized information and tourism services. Unfortunately, the most successful recommendation technologies (i.e., collaborative filtering and content-based filtering) used in practical settings, such as Amazon.com, cannot be easily adopted by the travel and tourism domain since, in particular for smaller tourism portals, no detailed purchase histories or preference profiles are available for the majority of users (Werthner and Ricci, 2004; Fesenmaier et al., 2006; Jannach et al., 2009). A further consumer- and demand-related research strand studies cause effect relationships between online satisfaction and the competitive performance of tourism enterprises and organisations. However, since experience- and emotion-oriented communication grew in importance as human-centric computing and emotionally intelligent interfaces emerged in tourism practice, the identification of the drivers behind online satisfaction and its impact on loyalty both to an organisation's programmes, services and web site has become an increasingly challenging and complex research task (Steinbauer and Werthner, 2007; Gretzel and Fesenmaier,

2009; p. 573). Moreover, despite virtual travel communities' large impact on electronic Word-of-Mouth as well as for the construction of memories and extended experiences in the post-consumption phase, research on this topic is still in its infancy (Pan et al., 2007; Buhalis and Law, 2008; p. 612; Vermeulen and Seegers, 2009). To conclude, by analysing content from virtual travel communities, tourism organisations can understand their customers' satisfaction and consumption behaviour and may undertake corrective actions for product improvements. Thus, Web 2.0 applications give consumers as well as tourism suppliers the opportunity to construct knowledge (i.e., business intelligence) about a destination or a tourism service from a great number of diverse sources (Fuchs and Höpken, 2009).

Although the Internet will continue to have a tremendous impact on the way consumers search for, purchase, consume and remember tourism experiences, recent technology developments aim at convergence, i.e., to gradually create one channel that satisfies all major information search, transaction and communication needs (Gretzel and Fesenmaier, 2009; p. 570). Accordingly, the research dimension *technological innovation* primarily aims at satisfying the requirements for effective integration of the entire range of hardware, software, groupware, netware and humanware (Werthner and Ricci, 2004). Selected examples for innovative technologies that blurred the boundaries between electronic equipment and software applications typically used in the travel and tourism industry are sketched next. To start with, dynamic *interoperable mechanisms* of collecting, processing and disseminating data within organisations and their extensive environment are increasingly based on alternatives to standardization, such as semantic web and ontology-based technologies, respectively (Höpken, 2000; Dell'Erba et al., 2005). Based on a domain-wide agreement of concepts and relationships between concepts (defined as ontologies) mediation mechanisms enable a translation between heterogeneous data formats of communication partners and a seamless data interchange. Secondly, since *wireless and mobile technologies* have become truly portable, wearable, wireless, global, integrated and smart (Gretzel and Fesenmaier, 2009; p. 573), they have created an increasing number of electronic connection points between customers and tourism suppliers. As a consequence, a growing vein of related e-Tourism research focuses on the multitude of location-based services (LBS) applied in a tourism context (Berger et al., 2003). LBS comprise the localisation of persons, objects and places, routing between them, search for objects in proximity (e.g., sights) and information about travelling conditions (e.g., traffic related data) (Höpken et al., 2006; Buhalis and Law, 2008; p. 616). A further research vein addresses both technical and social issues (e.g., technology acceptance, perceived information quality) with respect to *multimedia applications* in tourism. A recently launched example is Stubai Glacier's 3-D Map (2010) providing to travel planners almost real life impressions of the skiing experience through a virtual tour composed of aerial images and interactive digital maps. A final research area related to the technological innovation dimension is represented by *ambient intelligence* applications and their seamless inter-operation between different environments (Buhalis and O' Connor, 2005). By requiring contributions from various disciplines, '(…) advanced networking technologies and specific interfaces are aware of the characteristics of human presence, adapt to the needs of users, are capable to respond to indications of desire and result in systems that are capable of engaging in intelligent dialogue (…)' (Buhalis and Law, 2008; p. 617).

When discussing research topics related to the *industry function* it is crucial to remark that ICTS should be used to '(…) reengineer all business functions and processes towards supporting the organisation on its entirety rather than automating it (…)' (ibid., 2008; p. 617). Indeed, ICTS strengthen the relative competitiveness of (e.g., tourism) businesses by similarly affecting operational and strategic management. More precisely, according to Porter (2001) Internet technologies alter the barriers to entry (i.e., through creation of scale and scope effects and by increasing the amount of capital required), reduce switching costs, revolutionise distribution channels, enhance production efficiency and facilitate price-transparency, thereby generating transaction cost efficiencies and reducing uncertainty, complexity as well as information asymmetries. As a result, the customer turns out to be empowered through increased convenience, transparency, flexibility, direct communication, depth of available information and bargaining power through a greater range of available products. To summarize, the Internet supports the global proposition of tourism services for targeting niche markets of sufficient size, thus maximizing the value-added to individual consumers (Buhalis and Law, 2008; p. 618). However, with IT changing fast, competitive advantages never last long and travel and tourism industry has to constantly innovate in order to identify new value propositions and uses for emerging technologies (Gretzel and Fesenmaier, 2009; p. 572). Because of these trends, tourism players and e-Tourism researchers are similarly forced to rethink existing business models and take actions in re-developing tourism value chains (Buhalis and Law, 2008; p. 618).

The above discussion has demonstrated that the applied study field of e-Tourism research borrows from several scientific disciplines. For instance, by applying the elements of the broad spectrum of existing information technologies and tools (e.g., RFID, DSL, QR Codes, GPS, Java, XML, etc.) *computer sciences* provide insights into design and implementation principles of (web) applications and services relevant to travel and tourism industries. Furthermore, *information and behavioural sciences* translate particular requirements from the business sector or from final users into prototypically developed e-services (e.g., multi-functional platforms, augmented reality) and perform appropriate use-tests (e.g., technology acceptance, e-Business adoption behaviour). Moreover, by mutually adjusting business processes and business models, *management sciences* facilitate ICT adoption for operative (e.g., online auctions, revenue management, property management systems) and strategic purposes (e.g., business intelligence, smart business networks), respectively. Obviously, e-Tourism research is significantly informed by the *applied* and *cross-disciplinary study field of tourism research* covering both business-related (e.g., destination life-cycle, seasonality, airline, hospitality) and non-business-related issues (e.g., environmental aspects, quality of life). Finally, according to Tribe (1997), tourism research and related knowledge production processes are also affected from outside the typical disciplinary tradition (i.e., 'extra-disciplinarity'). More practically, although often highly contextualized for a specific project, the 'external world of tourism' (i.e., travel and tourism industry, government, think-thanks, interest groups and consultancies) developed a range of original ICT innovations, such as computerized reservation systems, yield management systems or smart hotel rooms (Tribe, 1997; p. 652). By summing-up the previous discussion, Figure 1. provides a multi-disciplinary framework of e-Tourism research.

Figure 1. A multi-disciplinary framework of e-Tourism research

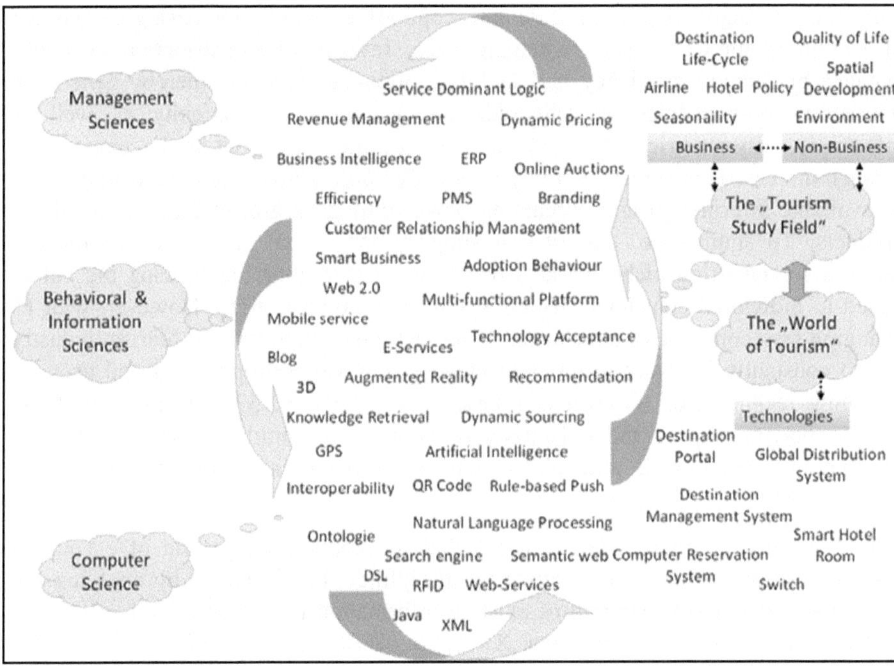

4 The mobile guide www.dolomitiSuperski.mobi

An exemplary application field strongly affecting tourists' trip behaviour (in particular during destination stay) is given by *mobile information services*. The benefits from *ubiquity* tourists may capitalize on stem from *spatial*, *temporal* and *contextual mobility* (Elgersson, 2006). Accordingly, tourists are in touch with the destination information network and can be reached at any place and any time using portable devices (e.g., mobile, smart-phone) (Fuchs and Höpken, 2008; p. 258; Rasinger et al., 2009). For instance, www.dolomitSuperski.mobi delivers tourists customized information regarding slopes and lifts, local and regional events, gastronomy and weather. Before application features and typical usage patterns are described in more detail, the technical framework upon the guide has been built-up is briefly sketched. Thus, since multiple research disciplines, like computer science (e.g., information systems development), information science (e.g., knowledge retrieval for business intelligence) and behavioural science (e.g., technology acceptance) are effectively blended, the presented case clearly demonstrates the multi-disciplinary nature of e-Tourism research

The architectural framework employed for mobile information systems provides basic components and services (e.g., data integration and persistence) as well as higher level business components (e.g., recommendation service, PUSH service) and enables the implementation of mobile applications by combining, configuring and extending the provided components. For instance, the recommendation service offers intelligent mechanisms to filter information according to the user requirements and profiles (Ricci and Werthner, 2001; Höpken et al., 2006), whereas the PUSH service sends information via SMS or email messages to final users and informs them about relevant events (e.g., changing weather conditions), tourism offers (e.g., cultural or sport events) or general circumstances (e.g., weather forecast or snow report). Concerning the information relevance, the push service takes into account the user profile (i.e., characteristics, preferences and usage history), the current context (i.e., time, location and circumstances) as well as the travel itinerary (i.e., already booked or selected tourism services or activities). Constraints (or rules) define which information is of relevance in which context and which tourism services are combinable with an existing travel itinerary. A user front-end enables a comfortable and flexible definition of rules (Beer et al., 2007).

Finally, the framework enables the flexible and dynamic adaptation of the user interfaces. Restrictions and capabilities of a corresponding device (e.g., display size, colours and pixel number) are explicitly considered in the adaptive visualisation process. Additionally, in order to provide comfortable and efficient access to tourism information via mobile devices to different user types, each customer segment is defined by customer profiles as well as usage patterns and may be individually addressed through a target group adequate design. Thus, the framework allows the application to be dynamically adapted with respect to content, presentation and application behaviour according to the used device, the target group and the actual context of use (e.g., time or location), respectively.

The framework has been instantiated for the skiing area DolomitiSuperski in the winter season 2007/08. Appropriate features and services have been selected from the overall framework and configured according to the specific requirements of DolomitiSuperski. Figure 2 shows the welcome page of the mobile application DolomitiSuperski.mobi and demonstrates one of the most powerful mechanisms of the framework – the dynamic adaptation to the used device (Höpken et al., 2008).

Figure 2. www.dolomitisuperski.mobi - PC Version versus Smart Phone

The left part of the figure shows the welcome dialog as it appears on a PC. The right part shows the same dialog automatically adapted to the specific capabilities and restrictions of a Smartphone. It has to be noted that the application itself has no knowledge on differing device specific presentations. Instead, this knowledge is completely externalized from the application code and represented by (XSLT-based) transformation instructions.

The mobile guide DolomitiSuperski.mobi shows the following features. In addition to some dozens of Livecam pictures weather information can be retrieved (i.e., mountain and valley temperature, mountain winds, one and three day forecast). The feature 'Lifts and Slopes' informs the user about length of open slopes within a specific skiing area, open ski lifts, snow type, snowfall at the mountain and valley as well as last snowfall. The 'SMS Service' provides a push-based slope report on snow quality and open slopes as well as a 'Good Morning Post' offering tips of the day and a weather forecast. Furthermore, the feature 'Dolomiti Ski Performance' allows the user to retrieve both mileage and differences in altitude left by skiing (i.e., discerned by the ski pass when entering the ski lifts). 'Dolomiti Dating' is a Web 2.0-based community platform which helps to find and to be found by others in order to jointly perform leisure activities in the various destination areas, such as skiing, cross-country, nightlife, meet and eat, etc. (Figure 3).

Figure 3. www.dolomitisuperski.mobi – Module: Dolomiti Dating

Further pull-based tourist information can be gathered for gastronomy, skiing huts, events of the day, tourist offices, ski pass offices, ski rental and ski schools. Moreover, 'Wallpapers' with impressive images from the destination zone may be downloaded on the mobile device. In addition to the above described application features offered all through an adaptive web application (as well as SMS-based PUSH service), the application DolomitiSuperski.mobi also provides an interactive ski map in form of a java-based application, installed on the mobile device of the user. Figure 4. shows the interface of the interactive ski map component, offering scalable maps which, based on hovering, show huts, mountain restaurants, slopes by difficulty degrees and lifts and cable cars.

Figure 4. www.dolomitisuperski.mobi – Module: Ski map

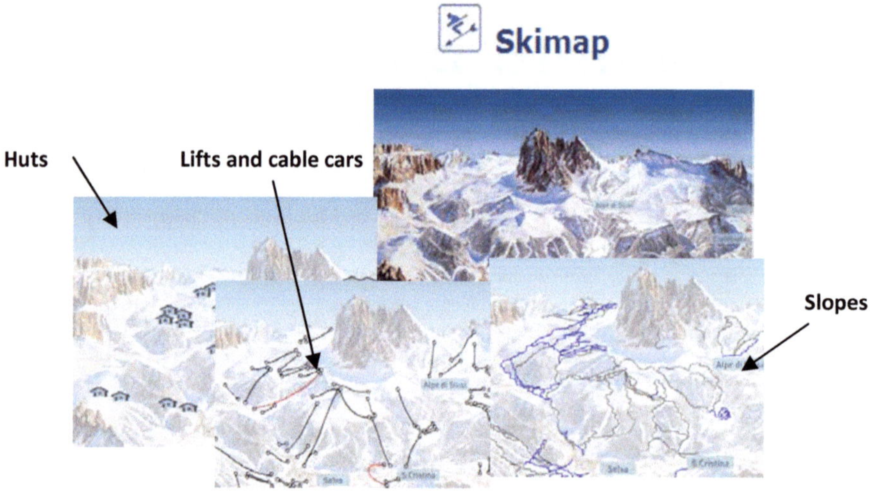

Already during the first implementation phase, the winter season 2007/08, at www.dolomitiSuperski.mobi a total number of 23,453 unique visits were registered on the log server (Figure 5). Interestingly enough, already one season after, in 2008/09 a total of 36,546 unique visits were registered. This is equivalent to an increase of usage rates at 56 % (Höpken and Fuchs, 2009).

Figure 5. DolomitiSuperski.mobi - Monthly usage statistics: Winter 2007/08 and 2008/09

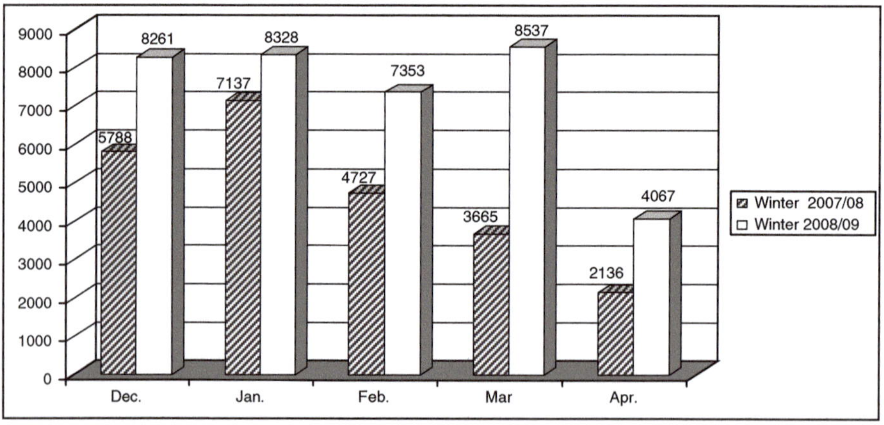

Web pages viewed by the users of the mobile tourist guide during winter season 2008/09 are typically registered in the form of log file data. In order to extract business intelligence the latter data has been used as input for both descriptive analysis and unsupervised learning methods performed with the data mining tool Rapid Miner™ version 4.2.

A look at the usage patterns of the application DolomitiSuperski.mobi shows that most accesses are registered at 13.00-14.00 hours and at 16.00-17.00 hours (i.e., the closing time of the ski area). Interestingly, again at 21.00 hours an increase in the usage statistics was observed. The daily usage patterns show that the application was most heavily used during New Year's Eve holidays 2009. A further descriptive analysis of log data revealed that the average session is composed of 5.39 page views with an average duration of 1.64 minutes. As expected for an Italian skiing destination, the largest user segment is represented by Italian users (37.5%), followed by German (18.6%), British (14.7%) and Czech (5.2%) users, respectively. The sum of sessions, the average number of viewed web-sites as well as usage duration of the mobile tourist guide is displayed per country of origin in Table 2.

Table 2. Sum of sessions, viewed web-sites and usage duration (Winter 2008/09)

Country of origin	Sum of sessions	Clicks per session	Duration (minutes)	Share %
Italy	7,413	6.15	1.80	37.5
Germany	3,691	5.01	1.53	18.6
United Kingdom	2,912	4.82	1.70	14.7
Czech Republic	1,045	5.45	1.66	5.2
United States	853	2.79	0.97	4.3
Poland	834	5.95	1.54	4.2
Switzerland	421	4.11	1.10	2.1
The Netherlands	417	6.88	1.92	2.1
Austria	416	4.57	1.35	2.1
Russian Fed.	320	5.67	1.65	1.6
Norway	214	5.02	1.63	1.0
Slovakia	210	8.01	2.31	1.0
Slovenia	189	5.21	1.46	0.9
Spain	174	7.71	2.24	0.8
Sweden	126	6.09	1.87	0.6
Denmark	122	5.80	1.93	0.6
Croatia	92	6.56	1.66	0.4
Israel	83	4.25	1.15	0.4
Ukraine	66	6.16	2.22	0.3
France	53	3.35	0.83	0.2
Japan	48	3.58	1.64	0.2
Finland	45	5.64	1.80	0.2

From Table 3 it emerges that the Livecam represents by far the most frequently requested feature of the tourist guide DolomitiSuperski.mobi (i.e., 27.1%). Moreover, four features show a rather equal level of attractiveness to the users, namely weather (i.e., 12%), lift and slope information (i.e., 11.6%), ski performance (11.1%) and the ski map feature (i.e., 9.1%). The mining methods most frequently applied in web usage mining are unsupervised learning techniques and association rule approaches (Fuchs and Höpken, 2009). Unsupervised learning techniques, such as clustering, are particularly used to discover homogenous user clusters (e.g., of sessions or transactions) or page clusters (Pitman et al., 2010). This is especially useful when segmenting e-commerce markets or for providing personalized web content for users with similar information interests. Thus, in a final attempt the *x-means* method was applied. As being a variation of the *k-means* method, it avoids the difficulty of having to fix the number of clusters beforehand (Hastie et al., 2009).

Table 3. Usage rates of DolomitiSuperski.mobi features: Winter 2008/09

DolomitiSupersi.mobi Feature	Clicks/ Feature	Share %
Livecam	17,562	27.1
Weather	7,757	12.0
Lift and Slope Info	7,540	11.6
Ski Performance	7,221	11.1
Ski Map	5,898	9.1
Wallpapers	4,149	6.4
Skiing Huts	2,643	4.0
Regional Information	2,511	3.8
Events	2,965	4.5
Gastronomy	1,920	2.9
Dolomiti Dating	1,453	2.2
SMS Service	1,256	1.9
Tourist Information	1,729	2.6

From Table 4 it emerges that seven clusters effectively portray homogenous user groups of the mobile guide DolimitiSuperski.mobi (Table 4). Interestingly enough, the majority of sessions (i.e., 94.1%) are triggered by four user segments showing highly concrete information interests. For instance, final users consolidated into cluster 1 (10.9%) are exclusively interested in Livecam pictures provided by DolomitiSuperski.mobi. Similarly, users behind cluster 4 (7.8%) and cluster 6 (13.1%) are solely concerned with weather information or using the interactive ski map, respectively. Rather unexpected, however, was the result with respect to cluster 5. Obviously, the majority (63.1%) is exclusively orientated towards hedonic features, such as using the Ski-Performance module reporting mileage and differences in altitude during a skiing day.

To sum-up, the application scenarios showed high usage rates standing well above 36,000 unique visits during the winter season 2008/09. Moreover, since the majority of users with concrete information interests have been effectively targeted by the mobile tourist guide, the modular configuration of the various application features looks rather optimal. Thus, the effectiveness of the previously presented architectural framework for developing mobile applications in tourism has been proved. To conclude, since the presented use case is based upon acceptance studies (Rasinger et al., 2007 and 2009), information systems development (Höpken et al., 2006) and knowledge retrieval for business intelligence (Fuchs and Höpken, 2009) it clearly demonstrates the multi-disciplinary nature of e-Tourism research studies.

Table 4. User clusters: DolomitiSuperski.mobi

	Cluster	Centroid Table	
	Cluster 1: Livecam		
Size (Sessions)	2,208	Livecam	3.37
Share (%)	10.9	Livecam Module Start	1.65
	Cluster 2: Lift and Slope Info, Livecam, Weather		
Size (Sessions)	433	Lift and Slope Info	6.41
Share (%)	2.1	Lift and Slope Module Start	2.51
		Livecam	1.09
		Weather	0.93
	Cluster 3: Huts, Livecam, Gastronomy		
Size (Sessions)	155	Region_huts	5.36
Share (%)	0.8	Huts	4.31
		Livecam	1.79
		Gastronomy	1.67
		Huts_Module Start	1.56
		Events	1.34
		Weather	1.26
		Lift and Slope Info	1.20
		Wallpaper	1.05
	Cluster 4: Weather		
Size (Sessions)	1,580	Weather	2.39
Share (%)	7.8	Weather Module Start	1.78
	Cluster 5: Performance		
Size (Sessions)	12,806	Ski Performance Module Start	1.43
Share (%)	63.1		
	Cluster 6: Ski Map		
Size (Sessions)	2,649	Ski Map	1.63
Share (%)	13.1	Ski Map Modul Start	1.39
	Cluster 7: Livecam, Lift and Slope Info, Weather		
Size (Sessions)	409	Livecam	10.86
Share (%)	2.1	Livecam Module Start	3.15
		Lift and Slope Info	1.61
		Weather	1.03

5 Conclusions and outlook

The research framework for e-Tourism developed in section three has highlighted the valuable relationship between computer, information and management sciences, respectively. Accordingly, effective e-Tourism practice and research will result from the interplay of information systems, technical and business processes as well as business models (Erl, 2005; Werthner, 2009). As displayed in Figure 6, these highly interrelated sub-systems may finally lead to innovative outputs, such as web services, e-services and customer services.

Figure 6. Extended service oriented architecture

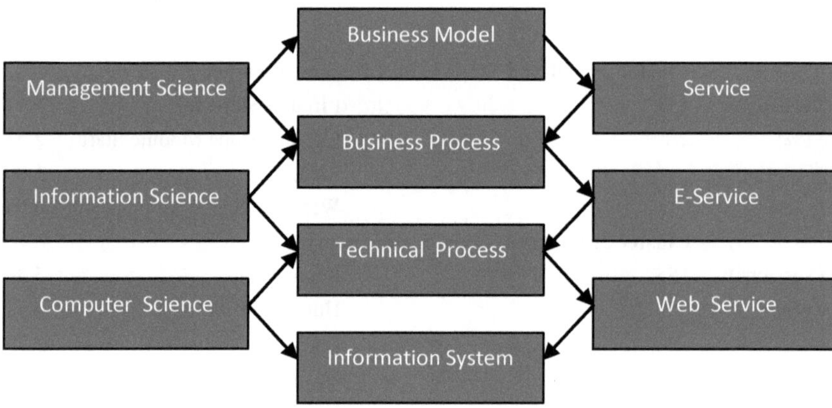

However, although the multi-disciplinary field of e-Tourism research has made substantial progress in the past 15 years, new ICT developments constantly demand new insights and understanding (Gretzel and Fesenmaier, 2009; p. 573). Actual research gaps in the field of e-Tourism are particularly apparent with regard to consumer experience and the use of emerging technologies, such as consumer generated media and virtual worlds as well as the consequences of IT use (Gretzel and Fesenmaier, 2009; p. 572). Thus, future e-Tourism research has to cope with the complex relationships between *technologies and organizational systems* (e.g., business and information systems, end-user applications, IT impact, web topologies), *data and information* (e.g., self-reinforcing propagation of user generated content, sentiment detection through content mining, etc.) and *users* (e.g., attitudes and behavior of tourists when using ICT applications, adoption and cooperation behavior of tourism suppliers). Indeed, what is needed is a better understanding of the organizational drivers behind IT adoption and the facilitators and inhibitors behind adjustment processes related to ICT innovations (Fuchs et al., 2009 and 2010). A best practice example for ICT innovations in tourism has been outlined in section four of this paper by discussing the architectural framework, application features as well as usage patterns related to the mobile tourist guide DolomitiSuperski.mobi.

From an industry point of view, an attempt was made to demonstrate that ICTS likely constitutes the biggest single force affecting structural and behavioural changes within travel and tourism industries. In particular, Web 2.0 applications and community sites suddenly created an empowered type of customer for which electronic worth of mouth becomes more important than any cleverly elaborated promotion campaign (Fuchs and Höpken, 2008; p. 259). Thus, final users' expectations for the 'ultimate' travel portal can be subsumed as being an unlimited online medium offering a broad spectrum of information generated electronically within the various (tourism) business networks and final user communities. Indeed, the future of e-Tourism will focus on consumer-driven technologies that support customers to dynamically interact with organisations along the tourism value chain. However, in order to generate real values from the many knowledge bases created by supply- and demand-side communities, new forms of lateral cross-supplier integration in electronic business networks become the crucial determinant of future business success (ibid., p. 260).

References

Beer, T., Fuchs, M., Höpken, W., Rasinger, J., and Werthner, H. (2007). CAIPS: A context-aware information push service in tourism. In Sigala, M., Mich, L., and Murphy, J. (Eds.), *Information and Communication Technologies in Tourism 2007* (pp. 129-140). New York: Springer.

Berger, S., Lehmann, H., and Lehner, F. (2003). Location-based services in the tourist industry. *Information Technology and Tourism, 5*(4), 243-256.

Buhalis, D., and O'Connor, P. (2005). Information and communication technology revolutionizing tourism. *Tourism Recreation Research, 30*(3), 7-16.

Buhalis, D., and Law, R. (2008). Progress in information technology and tourism management: 20 years on and 10 years after the internet – the state of e-tourism research. *Tourism Management, 29*(6), 609-623.

Compete, Inc. (2006). *Embracing consumer buzz creates measurement challenges for marketers.* Retrieved October 16, 2009, from http://media.competeinc.com/med/uploads/files/Compete_Spark_12_06_Embracing_Consumer_Buzz_Creates_Measurement_Challenges.pdf.

Conrady, R. (2007). *Travel technology in the era of web 2.0.* Berlin, Heidelberg: Springer.

Dell'Erba, M., Fodor, O., Höpken, W., and Werthner, H. (2005). Exploiting semantic web technologies for harmonizing e-markets. *Information Technology and Tourism, 7*(3/4), 201-219.

Dippelreiter, B., Grün, C., Pöttler, M., Seidel, I., Berger, H., Dittenbach, M., et al. (2007). online tourism communities on the path to web 2.0 - an evaluation, virtual communities in travel and tourism. *Information Technology and Tourism, 10*(4), 329-353.

eBusinessWatch (2006). *The European e-business market w@tch*, EC, Brussels: Enterprise and Industry Directorate General, Unit 4.

Elgersson, A. (2006). *The mobile destination - the use of mobile services in tourism destination development.* Göteborg: Göteborg University.

Erl, T. (2005). *Service Oriented Architecture – Concepts, Technology and Design.* NJ: Pearson Education.

Farrokh, M. (2009). Impact of E-Commerce on Travel and Tourism: A Historical Analysis, *International Journal of Management, 26*(3), 365-375.

Fesenmaier, D. R., Gretzel, U., Hwang, Y.H., and Wang, R. (2004). Applications of internet technology in travel and tourism. In Bidgoli, H. (Ed.), *The Internet Encyclopedia,* (Vol. 3, pp. 459-467). CA: Bakersfield, Wiley.

Fesenmaier, D. R., Werthner, H., and Woeber, K. (2006). *Destination recommendation systems: behavioural foundations and applications.* Wallingford: CABI Publishing.

Fuchs, M., and Höpken, W. (2008). Structural and behavioural changes on account of new information and communication technologies in tourism. In Kronenberg, C., Müller, S., Peters, M., Pikkemaat, B. and Weiermair, K. (Eds.), *Change Management in Tourism – From 'Old' to 'New' Tourism* (pp. 247-261). Berlin: Erich Schmidt Verlag.

Fuchs, M., and Höpken, W. (2009). Data mining im tourismus. *Praxis der Wirtschaftsinformatik, 270*(12), 73-81.

Fuchs, M., Scholochow, C., and Höpken, W. (2009). e-Business adoption, use and value creation – an Austrian hotel study, *Information Technology and Tourism, 11*(4) (in print).

Fuchs, M., Höpken, W., Föger, A. and Kunz, M. (2010). E-business readiness, intensity, and impact – an austrian destination management Organization Study. *Journal of Travel Research.*, (in print).

Forbes (2008). Retrieved January 14, 2010, from http://www.forbes.com/2008/09/26/fastest-growing-industries-lead-careers-cx_tw_0926jobgrowth.html.

Gretzel, U., and Yoo, K. H. (2008). Use and impact of online travel reviews. In O'Connor, P., Höpken, W., and Gretzel, U (Eds.), *Information and Communication Technologies in Tourism, ENTER 2008* (pp. 35-46). New York: Springer.

Gretzel, U., and Fesenmaier, D.R. (2009). Information technology: shaping the past, present and future of tourism. In Jamal, T., and Robinson, M. (Eds.), *The SAGE Handbook of Tourism Studies* (pp. 558-580). L.A.: SAGE.

Hastie, T., Tibshirani, R., and Friedman, J. (2009). *The elements of statistical learning – data mining, inference and prediction* (2nd ed.). New York: Springer.

Höpken, W. (2000). Reference model of an electronic tourism market. In Fesenmaier, D. R., Klein, S., and Buhalis, D. (Eds.), *Information and Communication Technologies in Tourism 2000* (pp. 265-274). New York: Springer.

Höpken, W., Fuchs, M., Zanker, M., Beer, T., Eybl, A., Flores, S., et al. (2006). Planner - an it framework for comprehensive and integrative travel guidance. In Hitz, M., Sigala, M. and Murphy, J. (Eds.), *Information and Communication Technologies in Tourism 2006* (pp. 125-133). New York: Springer.

Höpken, W., Scheuringer, M., Linke, D., and Fuchs, M. (2008). Context-based adaptation of ubiquitous web applications in tourism. In O'Connor, P., Höpken, W., and Gretzel, U. (Eds.), *Information and Communication Technologies in Tourism 2008* (pp. 533-544). New York: Springer.

Höpken, W., and Fuchs, M. (2009). Mobile Dienste für den Tourismus – Technischer Framework und exemplarische Instanziierung. In Egger, R., and Joos, M. (Eds.), *M-Tourism – Mobile Dienste im Tourismus*. Vienna: LIT Verlag.

Jannach, D., Zanker, M., and Fuchs, M. (2009). Constraint-based recommendation in tourism: a multi-perspective case study. *Information Technology and Tourism, 11*(2), 139-155.

Marcussen, C.H. (2009). *Trends in European Internet distribution of travel and tourism.* Retrieved September 28, 2009, from http://www.crt.dk/uk/staff/chm/trends.htm.

O'Connor, P. (2008). User-generated content and travel - a case study on Tripadvisor.Com. In O'Connor, P., Höpken, W., and Gretzel, U. (Eds.), *Information and Communication Technologies in Tourism 2008* (pp. 47-58). New York: Springer

Pan, B., MacLaurin, T., and Crotts, J.C. (2007). Travel blogs and the implications for destination marketing. *Journal of Travel Research, 46*(3), 35-45.

Pitman, A., Zanker, M., Fuchs, M., and Lexhagen, M. (2010). Web usage mining in tourism – a query term analysis and clustering approach. In Gretzel, U., Law, R., and Fuchs, M. (Eds.), *Information and Communication Technologies in Tourism 2010*. New York: Springer.

Porter, M. (2001). Strategy and the Internet. *Harvard Business Review 3*, 2–19.

Pyka, D. and Freitag, R. (2010). Overview of World Tourism 2009. In Conrady, R. and Buck, M. (Eds.), *Trends and Issues in Global Tourism 2010*, (pp. 3-10). Berlin: Springer.

Rasinger, J., Fuchs, M., and Höpken, W. (2007). Information search with mobile tourist guides - a survey of usage intention. *Information Technology and Tourism, 9*(3/4), 177-194.

Rasinger, J., Fuchs, M., Höpken, W., and Beer, T. (2009). Building a mobile tourist guide based on tourists' on-site information needs. *Tourism Analysis, 14*(3).

Ricci, F., and Werthner, H. (2001). Case base querying for travel planning recommendation. *Information Technology and Tourism, 4*(3/4), 215-226.

Sidali, K.L., Schulze, H., and Spiller, A. (2009). The impact of online reviews on the choice of holiday accommodations. In Höpken, W., Gretzel, U., and Law, R. (Eds.) *Information and Communication Technologies in Tourism, 2009* (pp. 87-98). New York: Springer.

Steinbauer, A., and Werthner, H. (2007). Consumer behaviour in e-tourism. In Sigala, M., Mich, L., and Murphy, J. (Eds.), *Information and Communication Technologies in Tourism, 2007* (pp.65-76). New York: Springer.

Stubai Glacier's 3-D Map. Retrieved January 25, 2010, from www.stubaiergletscher.com/en/winter/ panorama/.

Tribe, J. (1997). The indiscipline of tourism. *Annals of Tourism Research, 24*(3), 638-657.

UNWTO (2009). *World tourism barometer.* Retrieved January 14, 2010, from http://unwto.org/facts/eng/barometer.htm.

Vermeulen, I.E., and Seegers, D. (2009). Tried and tested: the impact of online hotel reviews on consumer consideration. *Tourism Management, 30*(2), 123-127.

Werthner, H. (2007). E-tourism in Austria – a research view, Lecture held at '*Austrian eTourism Day*', January 2007, Tech Gate, Vienna.

Werthner, H. (2009). *Technological challenges for destination management systems*. Key Note Speech held at ENTER@2009, Amsterdam, Netherlands.

Werthner, H., and Ricci, F. (2004). E-commerce and tourism. *Communications of the ACM, 47*(12), 101-105.

Xiang, Z., and Gretzel, U. (2009). Role of social media in online travel information search. *Tourism Management 31*(2), 179-188.

Zanker, M., Jessenitschnig, M., and Fuchs, M. (2009). Automated semantic annotation of tourism resources based on geo-spatial data. *Information Technology and Tourism, 11*(3).

Zhu, K., Kraemer, L., and Dedrick, K. (2005). IT payoff in e-business environment: an international perspective on value creation. *Journal of Management Information Systems, 21*(1), 17-54.

The Risk™ strategic game of rural tourism: how sensory analysis can help in achieving a sustainable competitive advantage

Cristina Santini[a]
Alessio Cavicchi[b]
Maurizio Canavari[c]

[a] Università Telematica Internazionale - Unitel
santini.cristina@gmail.com

[b] University of Macerata
Department of Studies on Economic Development
a.cavicchi@unimc.it

[c] Alma Mater Studiorum – University of Bologna
Department of Agricultural Economics and Engineering (DEIAgra)
maurizio.canavari@unibo.it

Abstract

This paper examines how sensory analysis can be effectively implemented to achieve competitive advantage in rural tourism. Starting from the theoretical framework of experience economy outlined by Pine and Gilmore (1998), the authors describe how firms can differentiate their offer by using sensory analysis. In the first section, the authors provide a description of the rural tourism phenomenon and a literature review of rural tourism. Then, descriptive examples of the implementation of sensory analysis in rural tourism are provided. The paper ends with conclusions and recommendations for rural tourism operators.

Zusammenfassung

Dieser Beitrag untersucht, wie die sensorische Analyse effektiv genutzt werden kann, um einen Wettbewerbsvorteil im Agrotourismus zu erzielen. Von dem theoretischen Rahmen der von Pine and Gilmore (1998) beschriebenen *experience economy* ausgehend wird dargestellt, wie Unternehmen ihre Angebote durch sensorische Analyse differenzieren können. Im ersten Abschnitt wird das Phänomen „Agrotourismus" beschrieben. Anschließend werden Beispiele der Anwendung sensorischer Analysen im Agrotourismus vorgestellt. Der Beitrag endet mit Schlussfolgerungen und Empfehlungen für Agrotourismus-Veranstalter.

Riassunto

Questo paper descrive come l'analisi sensoriale può essere efficacemente impiegata per il conseguimento di un vantaggio competitivo nell'ambito del turismo rurale. Partendo dal modello teorico dell'experience economy delineato da Pine and Gilmore (1998), gli autori descrivono come le imprese possono differenziare la propria offerta attraverso l'utilizzo dell'analisi sensoriale. Nella prima sezione del lavoro gli autori propongono una descrizione

K.L. Sidali et al. (eds.), *Food, Agri-Culture and Tourism: Linking local gastronomy and rural tourism: interdisciplinary perspectives*, DOI 10.1007/978-3-642-11361-1_11,

del fenomeno del turismo rurale e una rassegna della letteratura sul tema. Successivamente vengono illustrati esempi su come l'analisi sensoriale può essere utilizzata nel turismo rurale. Nella sezione finale del paper gli autori presentano le conclusioni e dei suggerimenti per gli operatori del settore.

Keywords: Rural tourism; Sensory analysis; Differentiation strategy.

1 Introduction

Business practice suggests that when a business becomes profitable, the number of competitors rises. Surviving fierce competition requires from entrepreneurs a strategic mindset, and when playing the famous strategic board game Risk™, you have to acquire the proper set of resources if you want to succeed against your opponents. Resources must be inimitable and unique to gain a sustainable competitive advantage (Barney, 2001a; Barney, 2001b; Porter, 2008; Wernerfelt, 1984). How can theory be translated into practice and be applied to rural tourism? Or, in other words 'what makes a tourist product/service unique?' During recent years the businesses of tourism and high quality food and wine has experienced a dramatic change: the number of food and wine quality certifications has increased as well as the demand for quality and *terroir* food products and tourist destinations linked to food and wine. Attracted by profits and by the availability of new customers, many firms have jumped into the business of quality and typical food products, resulting in a proliferation of products on the market and an increase in new producers and local associations. As long as market conditions have been favorable, firms have imitated each other in strategy making. As competition became fiercer, companies had to rethink their strategies in order to defend their competitive positioning or to create a value innovation (Kim and Mauborgne, 2004; 2005). In this perspective, sensory analysis can play a significant role in defining a winning strategy.

In particular this paper wants to answer the following research question: what is the role of sensory analysis in promoting the uniqueness of a tourist product? How can sensory analysis make a tourist experience unique? In order to answer these research questions, after describing the rural tourism phenomenon in Italy, the theoretical framework of *experience economy* outlined by Pine and Gilmore (1998) will be described. A literature review will be carried out in order to explain the rural tourism phenomenon. Then descriptive examples of the implementation of sensory analysis in rural tourism will be offered. We conclude with recommendations for rural tourism operators.

2 Rural tourism in Italy: a sleeping giant

According to the definition provided by the OECD (1994) Rural Tourism can be defined as "*tourism happening in rural areas*". As has been recently underlined by Ecorys (2009), it is difficult to estimate rural tourism trends: official statistics do not include data about small scale accommodation in rural areas. Nevertheless, it is estimated that in Europe the bed capacity of small, micro or complementary accommodation in rural areas accounts for 6.5 million bed places, corresponding to more than 500,000 accommodation units. Among them, "agritourism" represents about 15-20% of the total. According to Eurogites, the European Federation for Farm and Village Tourism, in the past 15 years agritourism has registered an annual growth of 10-15% per year both in demand and supply, which is three times the growth shown by European tourism in general (retrieved November 10, 2009, from http://www.eurogites.org). It seems appropriate to define rural tourism as a 'sleeping giant', although the size of this phenomenon is difficult to measure exactly also in Italy (Ecorys, 2009). A fragmentation on the supply side and a proliferation of bottom-up initiatives promoted by many associations for rural tourism development indicate increasing opportunities for the rural tourism industry.

In the following section we focus on some aspects of rural tourism, and, more specifically, on agritourism and oeno-gastronomical tourism.

2.1 Agritourism

First of all, we must highlight that rural tourism does not exactly correspond to agritourism: according to statewide Small Farm Program of the University of California, agritourism can be conceived as a subset of the wider industry called rural tourism which also includes resorts, bed-and-breakfast accommodation in private bedrooms, off-site farmers' markets, non-profit agricultural tours, and other leisure and hospitality businesses that attract visitors to the countryside (retrieved November 10, 2009, from http://www.sfc.ucdavis.edu/agri tourism). Thus, rural tourism differs from agritourism in two ways: 1) rural tourism enterprises do not necessarily have to be managed by farmers, and 2) rural tourism activities are not necessarily aimed at generating supplemental income for farmers.

Nevertheless, agritourism is carefully monitored and the data on its development can be considered a good proxy for understanding the broader movement of rural tourism. Concerning agritourism in Italy, data is provided by ISTAT (Italian National Institute of Statistics) and by other non-profit associations or entrepreneurial lobby groups. According to ISTAT, over the ten years from 1998 to 2007, the number of agritourism enterprises grew from 9,700 to 17,700 units (+82.3%). The total turnover of agritourism is estimated at 200 million Euros (Ecotur, 2008). Among the new agritourism operations, those providing food and wine-tasting opportunities have increased by 193.5%. Currently there are 2,664 firms. Three leading regions host most agritourism operations: Tuscany (4,100 enterprises), Trentino (3,270) and Veneto (1,218).

Rural tourism and agritourism have universally registered positive performance in developed countries. In Italy rural tourism and agritourism can be considered as facets of the so-called 'Made in Italy' (Italian style) phenomenon which has been so successful that it cannot be classified as a niche market anymore (Esposti, 2006). It would be extremely useful to investigate relationships between rural tourism, agritourism and food-and-wine businesses in order to define and evaluate growth paths.

2.2 A business around food and wine

In 2007, 4 million wine tourists created in Italy a turnover accounting for 2.5 billion Euros (Somoza, 2007). The phenomenon seems even greater if we consider tourism linked to food (food tourism): in Italy 30 million Italian tourists and 20 million foreign tourists have visited, at least once a year, a place in order to enjoy its food or wines. The total turnover (food and wine) in 2007 was 5 billion Euros.

In order to promote wine tourism many new local associations have been established: in Italy there are 120 wine routes and 546 wine cities (Osservatorio Turismo del Vino, retrieved November 10, 2009, from http://www.cittadelvino.it), 284 olive oil cities (*città dell'olio*, retrieved November 10, 2009, from http://www.cittadellolio.it), 60 organic farming cities (*città del biologico*, retrieved November 10, 2009, from http://www.cittadelbio.it). 42 bread cities (*città del pane*, retrieved November 10, 2009, from http://www.cittadelpane.it). Local food production brings to the market about 4,400 traditional food products among which we can find 357 quality wines[1] produced in designated regions plus 120 wines with a typical geographical indication, as well as 183 protected designations of origin (PDO) and protected geographical indications (PGI) for food products.

The growing demand for rural tourism destinations has induced an expansion in the typical product supply. Attracted by profits and by the availability of new customers, many firms have jumped into the business of quality and typical food products resulting in a proliferation of products on the market and the appearance of new producers and local associations.

As a result, supply is extremely fragmented and achieving effective differentiation becomes tremendously hard. In many cases the excessive proliferation of appellations and geographical indications creates confusion in the consumer's mind: "consumer confusion" has been defined as an uncomfortable state of mind that primarily arises in the pre-purchase phase and which negatively affects consumers' information-processing and decision-making abilities and can lead to consumers making suboptimal choices (Walsh, 1999).

2.3 The main actors in rural tourism in Italy

The phenomenon described above is promoted by several associations that work to improve networks and awareness and promote rural destinations. Three of them are a

---

[1] Four major categories are established by the Italian quality regulations for wine: Vino da Tavola (table wine) indicates any wine without the indication of geographical provenance, Vino a Indicazione Geografica Tipica (IGT) indicates a table wine produced in a specific area, Vino a Denominazione di Origine Controllata (DOC) indicates that high quality wines are produced in specific well-defined regions, according to specific rules designed to preserve the traditional wine-making practices of the individual regions. Vino a Denominazione di Origine Controllata e Garantita (DOCG) is a more stringent regulation than DOC and it requires that the wine is evaluated by a tasting committee before it is bottled. This wine classification is presently under revision to harmonize with the EU regulation on the protection of Geographical Indications and Designations of Origin (Council Regulation (EC) No 510/2006).

sort of spin-off of traditional Italian farmers organizations unified under the national *aegis* of Anagritur.

Terranostra (Coldiretti) is a national association that groups 18 regional and 96 provincial associations; through the guide for rural tourism published by the association, Terranostra promotes tourism in specific areas. The association also supports marketing initiatives and development projects at a local level, such as the publication of specific local guides, training sessions, development projects, and partnerships with schools for educating students regarding agritourism.

Turismo Verde is an association created in 1981 by the Italian agricultural confederation that promotes hundreds of farms through its guide 'agriturismo in italia'. Turismo Verde also has some regional websites for promoting educational farms and 'open farms' operations. The association also publishes books and articles and organizes meetings for supporting startups in the agritourism business.

Agriturist is the oldest rural tourism association, created in 1965, promoting around 200 farms. Agriturist promotes rural tourism through the offer of 'green holidays packages' and through the 'guide to rural hospitality', which was published for the first time more than 30 years ago, and through the more recent "guide to farm products" (retrieved November 10, 2009, from http://www.ruraltourism international.org/Italy).

Slow Food is a non-profit, member-supported eco-gastronomic organization which was founded in 1989 to counteract fast food and fast life, the disappearance of local food traditions and people's dwindling interest in the food they eat, where it comes from, how it tastes and how food choices affect the rest of the world.

Wine routes or ***Strade del Vino*** became widespread in Italy. Gatti and Incerti (1998) define a wine route as a kind of cultural itinerary; along a sign–posted itinerary visitors can discover products and wines in a certain region by visiting wineries. Once a wine route is established in a certain area, wineries can decide to join the wine route network. As well as wine routes, the ***Strade dei Sapori*** – which literally means flavor routes – provide tourists with a gastronomic itinerary. In some areas, under the aegis of regional administration, the ***Strade del Vino*** and the ***Strade dei Sapori*** are jointly promoted.

The ***Movimento Turismo del Vino*** (Wine Tourism Movement) was established in 1993 with the aim of encouraging "Italian wineries to open their doors to visitors". The main event is "Cantine Aperte" (open cellars), which is organised every year in Italy during the last Sunday in May. The wine cities association ***Città del Vino*** was established in 1987 and operates at a national level. The main duties of the association are promoting and improving tourism and cultural resources, with particular attention to the local and traditional food products and quality wines of its 569 member Municipalities.

Two more specific associations have been established for the promotion of olive oil and bread. The ***Città del Pane*** association – bread cities – includes more than 40 cities which are specialized in bread production; the ***Città dell'Olio*** association – olive oil association – includes 341 cities famous for their olive oil production. The aims of these associations are to educate tourists and consumers about bread and olive oil production and consumption by organizing events, implementing marketing and communication strategies and promoting local tourism.

Finally, one must remember the role played by the **World Wildlife Fund** (***WWF***) in promoting and supporting sustainable rural tourism. Besides the associations that operate at a national level hundreds of smaller associations, whose aims are the promotion of rural tourism and the value-enhancement of local products, operate at a local level.

3 Rural tourism as leverage for rural development

The academic debate about rural tourism is extremely lively and intense. Rural tourism can be conceived as a flywheel for the economic development of rural areas and its populations (Williams and Shaw, 1998). Nevertheless, one must not forget the role that rural tourism can play in social agriculture (Carbone, Gaito and Senni, 2005). The World Tourism Organisation (WTO) seminar on rural tourism in Europe (WTO, 2003) clearly specifies some interesting aspects of rural tourism: it is an alternative to mass markets; it spreads a sustainable approach to economic development; it helps in creating private/public partnerships; it provides the opportunity for harmonizing standards for the services offered; it offers the chance to some key groups to be more involved in economic societies.

The problem-solving orientation in rural tourism should also be remarked on: rural tourism can help in increasing economic income in rural areas (Hummelbrunner and Miglbauer, 1994) and has been an optimal solution for rural areas encountering problems (Gannon, 1994; OECD, 1994). Interest in the role of rural tourism in economic development has been proven by extensive research available in this field: some authors investigated whether rural tourism could provide effective leverage for economic development (Briedenhann and Wickens, 2004; Greffe, 1994), whilst others underlined specific issues of rural tourism, such as sustainability (Middleton and Hawkins, 1998; Sims, 2009), or described the development of rural tourism in some specific areas (Briedenhann and Wickens, 2004; Fleischer and Pizam, 1997; Reichel, Lowengart and Milman, 2000).

The strategic facet of rural tourism is extremely interesting: developing rural tourism has been the strategy adopted by some areas to diversify their tourism offer. As outlined during the WTO seminar (WTO, 2003), rural tourism attracts specialized niche tourists. In recent years firms settled in rural areas have understood the potential of rural tourism, also thanks to the development policies run by the European Union, with particular reference to the LEADER and LEADER+ programs which are managed at a local level by individual Regional administrations through specific rural development programs.

Firms cry out for differentiation: if in the past differentiation was a generic strategy (Porter, 1985; 1998), nowadays it is seen as the only way to defend a firm's competitive position through creating or discovering new niche markets where competition is lower (Kim and Mauborgne, 2000; Kim and Mauborgne, 2007). However, first of all these firms have to deal with lack of professionalism:

> "Many of these alternative accommodations – especially in mainly rural areas – once started as a strategy to raise income and the entrepreneurs often have no specific educational background in business management or tourism. As hotels are being taken over by a next generation of entrepreneurs, it is expected that professionalism will increase" (Ecorys, 2009; p. 57).

The role of the tourist in rural tourism should also be emphasized: "Rural tourism is engaged in by tourists seeking rural peace; it is tourism away from the mainstream, away from areas of intensive tourism activity. It is engaged in by visitors who wish to interact with the rural environment and the host community, in a meaningful and authentic way" (WTO, 2003). Background research has outlined how rural tourism responds to a broad range of motivations in tourists which go from the need to try out ecological uniqueness to the wish to enjoy a countryside experience (Page and Getz, 1997; Sharpley and Sharpley, 1997).

In conclusion we can say that firms involved in rural tourism must take into consideration the following issues:

- There is growing competition in the rural tourism business due to the massive entrance of firms into the business;
- There is a need for training of employees in order to respond to the lack of professionalism in the small scale accommodation business;
- There is a pressing need to understand tourists desires and expectations in order to implement effective differentiation and rethink companies' strategies: as long as the market conditions have been favorable, firms have adopted an imitative behavior in strategy making or have marketed their products by focusing on the reputation of the surrounding area, but now companies need to understand under which conditions the service/product they offer may be defined as unique.

4 Food tourism and the role of *terroir* products

Garrod, Wornell and Youell (2006) illustrate the role of countryside capital (landscape, biodiversity, historical features, distinctive local customs and way of life, etc.) in the development of sustainable rural tourism. The authors emphasize how rural tourism businesses and destinations exploit countryside resources to attract tourists and offer them fulfilling experiences during their journey.

Aurier, Fort and Sirieix (2005) identified three dimensions, cognitively describing terroir products in the consumer's mind:

- Trade-skill, including the categories know-how, recipes and tradition
- Time and culture, including the categories history and ritual
- Origin, including the categories territory, region and land

Figure 1. Interaction between rural tourism and countryside capital

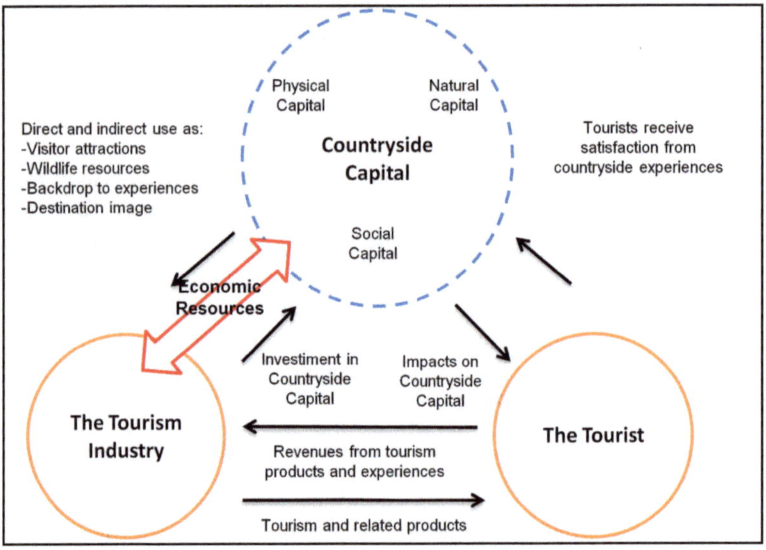

Source: Garrod et al. (2006)

Terroir products can be conceived as a synthesis of production processes based on local resources which directly link, case by case, though in different ways, the product to the territory. In this way, the product is a harmonious bundle of territorial, climatic, historical and cultural characteristics which belong to a certain region (Arfini, 2005). This kind of product becomes the cultural marker of a territory and the identity and image of this product provide leverage for the development of the general welfare of the population living in that specific region.

As underlined by Bessière (1998), culinary heritage may be used as a means to boost development and numerous communities. In Italy, the perception that the territory can be revitalized using cultural values and identity through encouraging local operators to promote transmitted skills and expertise is gradually strengthening. Moreover:

> "Heritage valorization represents not only a fashionable trend but also social, economic and political determination. Indeed, if local authorities are experiencing an identity crisis and are discovering a new-found interest for their own gastronomy, heritage promotion may turn out to be profitable. As tourists come into their area, they buy local products or eat in restaurants or on farms, thus representing a sizeable source of income for local communities." (Bessière, 1998; p. 32)

Recently, Hall et al. (2003) talk for the first time about a certain "tourism of taste". They summarize the importance of studying food tourism as follows: a) food is an attraction in its own right to travel, for instance in visiting a specific event or a firm like a winery or a brewery, b) a particular destination could have a great reputation for certain products, i.e., wine from Tuscany could be an example of special interest tourism (Hall and Mitchell, 2001), c) being part of tourist experience, food has become an important element in the marketing of tourism and in determining visitor satisfaction.

Figure 2. Food tourism

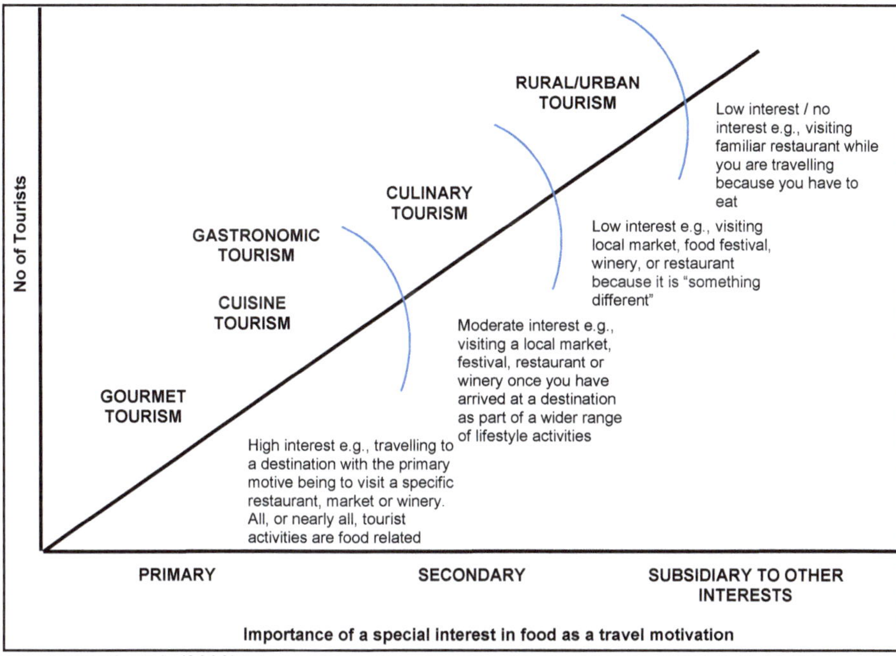

Source: Hall et al. (2003)

Food tourism can be classified into different types of tourism, according to the interest tourists have in food as a travel motivation and according to the number of tourists served. Gourmet tourism and rural tourism lie at the opposite end of the scheme provided by Hall et al. (2003). Gourmet tourism addresses its offer to a small number of selected tourists, whose main interest for travelling is tasting specialty foods or wines in selected restaurants or wineries. On the contrary, the phenomenon of rural tourism involves a wider number of tourists whose interest in food could be secondary or subsidiary to other interests: this is the case with tourists that visit food and wine festivals while visiting a certain rural area. This scheme summarizes the many facets of food tourism and describes the relationships that rural tourism has with food; furthermore it provides a clear idea of how differentiation in food tourism can happen.

5 The role of sensory analysis in the improvement of rural business experience economy

Paraphrasing the seminal article by Pine and Gilmore (1998) on "experience economy", a firm (or a group of firms, consortia, etc.) working in the rural tourism sector, before charging admission, should design an experience that customers judge to be worth the price. As shown in figure 3, two dimensions (graphically represented as two axes) are necessary to develop great experiences: customer participation and environmental relationships. The first axis has on one side the concept of "passive participation" where customers do not affect the reality at all, whilst on the other side lies "active participation" in which the tourist has a prominent role in designing the event. The second axis depicts "external relationship" and has at its two extremes "absorption" and "immersion". These two concepts have opposite meanings regarding the level of involvement of the tourist with the context. The spectra of the two dimensions define four categories of experiences: *entertainment* (in which tourists participate more passively than actively, and the connection with the reality in front of them is more of absorption than of immersion), *education* (when customers are more actively participating but not immersed in the action), *escapist* (where people are more involved than in educational or entertainment experience) and finally, in the case of less personal participation of tourists, *esthetic*.

Figure 3. The four realms of experience

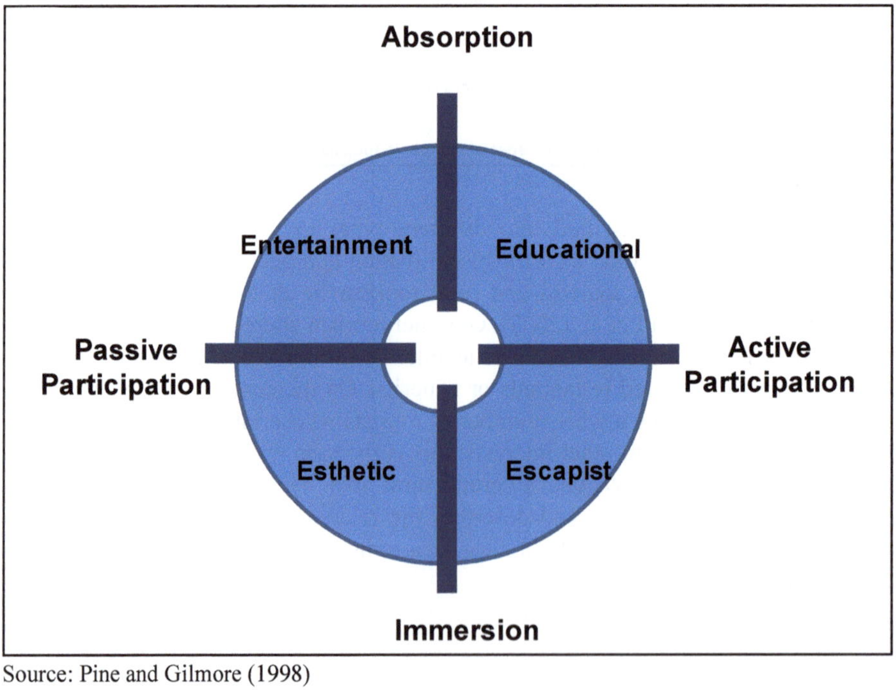

Source: Pine and Gilmore (1998)

The recipe provided by the authors to design memorable experiences entails five steps:

(a) theme the experience,
(b) harmonize impressions with positive cues,
(c) eliminate negative cues,
(d) mix in memorabilia,
(e) engage all five senses.

In summary they can be explained as follows: the theme chosen should conduct the whole experience toward a unified story able to catch the customer attention, involvement and participation through positive and consistent (with the theme) cues that create impressions. Impressions are defined by the authors as the "takeaways of the experience". Thus, each cue or detail needs to be coherent with the general theme given to the experience. In order to do this, it is necessary to avoid every negative cue that might diminish, contradict or distract from the theme. Then, the availability of memorabilia such as postcards, t-shirts or other physical reminders of the experience can be important to evoke the impressions. Probably, the most important step to emphasize the magnitude of the experience is the engagement of the five senses. The authors stated that "the more senses an experience engages, the more effective and memorable it can be. Not all sensations are good ones, and some combinations don't work". Smells, flavors, noises need to be deeply studied to verify any coherence among them and any leaning towards the theme chosen by the local operators. This leads to sensory analysis. Accordingly, we will show how this can be implemented in the chosen model. Nevertheless, the expression 'sensory analysis' is over cited. In fact, nowadays, many promotional and communicational activities in both the fields of 'food and wine business' and 'tourism sector' are explained using the expression 'sensory analysis'.

5.1 Sensory analysis: definitions

What is sensory analysis? According to Drake (2007), sensory science can be traced back to the 1800s, with the development of psychological theories to measure and predict human responses to external stimuli (Lawless and Heymann, 1999).

In a recent book edited by Moskowitz, Beckley and Resurreccion (2006), two definitions of sensory evaluation are offered: a) the first is proposed by the Institute of Food Technologists and enunciates that "Sensory evaluation is the scientific discipline used to evoke, measure, analyze, and interpret reactions to the characteristics of food and materials as they are perceived by the senses of sight, smell, taste, touch, and hearing" (retrieved November 10, 2009 from http://www.ift.org/divisions/sensory) b) the second has been elaborated by Herbert Meiselman in 1993: "Sensory evaluation is a science of measurement. Like other analytical test procedures, sensory evaluation is concerned with precision, accuracy, sensitivity and avoiding false positive results" (Moskowitz et al., 2006).

From both definitions it is clear that we are dealing with a science in which statistics cover a prominent role: as a demonstration of this, a new scientific society called the 'sensometric society' has been founded by many researchers belonging to this field of study [2]. Many sensory tests exist and they can be applied to answer to a bundle of questions about consumers' perception of food and non-food products.

5.2 Sensory analysis: techniques

Currently, sensory analysis can be categorized into three basic categories or groups of tests (Drake, 2007): traditional, analytical and affective (consumer) tests.

With the techniques belonging to the first group an overall quality score is assigned to a product based on a designated list of defects. These specific tests are currently used mainly for quality controlling. The easiness and convenience of these tests make them suitable for large manufacturers or quality control systems that need quick decisions.

The second group of tests uses panelists to define the differences among products. Panelists need to be trained and training length depends on the complexity and the number of sensory attributes to be profiled: training can go from a few to hundreds of hours in order to make panelists operate in unison like musical instruments within an orchestra. Each individual panelist serves a function analogous to an individual sensor on an instrument. In this second group we can find:

- Discrimination tests such as triangle and duo-trio tests: they are used for underlining and grading differences among products;
- Threshold tests: these define the lowest threshold with a minimum concentration of a certain characteristic that can be perceived; they are particularly useful for defining the level of desirable and undesirable presence of components in food;
- Descriptive analysis: this consists of training a group of individuals to identify and quantify sensory attributes of a product.

The third group of tests is performed directly with consumers. Consumer tests, both quantitative and qualitative are useful to understand likes and dislikes, preference and acceptance with regard to a certain product. What is important here is the 'voice' of the consumer, this is particularly indicated when a product development process needs to be carefully performed.

It is clear that most of the time we hear about sensory analysis in tourism programs or in other promotional events, it is not 'sensory analysis' as a science being mentioned. There is a sort of distortion which brings the tourist/consumer to a mix of sensory impressions. On the other hand, the application of scientific standards, such as in the case of descriptive analysis, could create the basis for a 'shared language' – made of objective sensory properties – that consumers could use when talking about a product; furthermore this 'shared language' could be useful for enhancing fascination in

[2] The aim of this scientific society is to increase awareness about the need that sensory and consumer science has for special methodology and statistical methods (www.sensometric.org). The official journal of the society is *Food Quality and Preference*. For further information about the scope and methods of sensory analysis see the Journal website on *www.elsevier.com*

consumers' minds and could be effectively used for communication and promotional events.

Here we propose the example of the sensory profile of five different Malvasia wines produced in 5 different areas (Lanza, Mazzaglia and Pagliarini, 2009). Figure 4 shows some common attributes, elicited by trained panelists, that differ in intensity depending on the peculiarity of the harvesting area. In the article we read:

> "Malvasia cultivars distributed in the Mediterranean basin for more than 500 years have been modeled by pedoclimatic environments and by growing techniques, connected to multiple cultural factors, that each population has developed with time. Such interaction has led to products with sensory characteristics which are qualitatively and quantitatively different from each other, contributing to the expansion of the range of Malvasia wines nowadays available" (Lanza et al., 2009; p. 6).

Figure 4. Sensory profile of the Mediterranean Malvasia wines

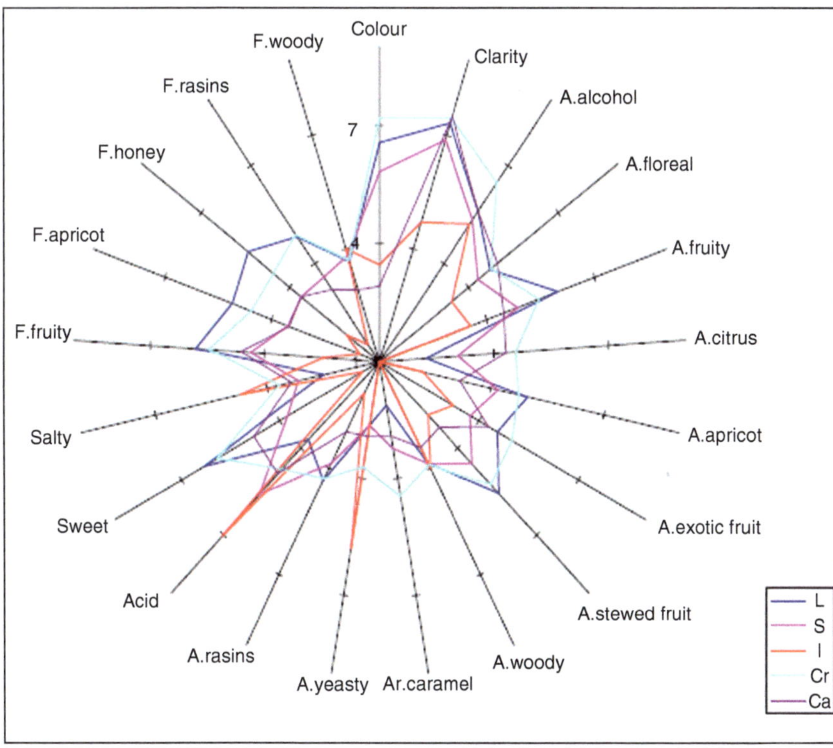

Note: L = Lipari; S = Sardinia; Ca= Canaries; Cr = Crete; I = Istria

Source: Lanza et al. (2009)

This sort of “sensory identity card” allows the consumers and producers to talk a common language where, as highlighted by Pine and Gilmore (1998), impressions lie in clear cues which are coherent with and related to territorial features.

Through practical exercises based on sensory tests techniques, tourists can be trained to understand the distinctiveness of the territory. The impressions induced by this exercise become memorabilia and can be taken back home (along with other gadgets) together with the ‘identity card’ of a certain product.

6 Experience economy and sensory analysis: some evidence from the rural tourism sector

Fiore et al. (2007), elaborating on the original work by Pine and Gilmore (1998), formulate a tool for small rural businesses who wish to enrich consumer experiences. They suggest including all four types of experiences (4Es) for creating an optimal rural tourism event. However it is reasonable to focus initially on one or two of the 4Es and then augment their experiential offerings over time.

Fiore et al. (2007) propose some examples in rural tourism that match the definition of Pine and Gilmore. For instance the *educational* experience can be found when a bakery enables customers to learn about the production of ornately decorated cookies by watching and then trying decoration techniques; the *esthetic* experience could be the Victorian garden of a bed-and-breakfast that enraptures guests with colors, scents, and textures; local farmers drawing spectators/tourists through sponsored piglet races at a town festival could represent the *entertainment* experience. The *escapist* experience for restaurant visitors could be represented by the Italian employees who emphasize the proper Italian pronunciation of menu items. The examples provided are numerous and they are undoubtedly useful suggestions for rural businesses.

In the following, we stress the importance of sensory analysis as a technique to improve visitors’ experience.

Due to its scientific nature, sensory science becomes an important tool for *educational experience*. In her recent book entitled “The taste of place: a cultural journey into terroir”, Amy Trubek (2008) explores the connections between taste and place, examining culinary realities in France and the US. She ably explains the importance of sensory analysis in educational experience:

> “One day I attended the annual Exposition of the Coteaux du Languedoc, an event championing a relatively new AOC wine-growing region representing more than a hundred wineries… The event was held in the Mas (an Occitan term for farm) de Saporta… Of the series of tastings held in this room during the exposition I chose to attend “Eveil des Arômes pour les Enfants”. An aroma tasting for children? At a wine expo? I was intrigued. This was a first for me, perhaps because you would never find such an event in the United States, for beyond our notion that kids love sweets, we do not really consider children capable of discerning taste […]. The entire event was patterned on a traditional

> wine tasting, but with fruit juice instead of wine. The tasting reveals a fundamental assumption of the French foodview: when you eat or drink, it needs to be a shared experience that incorporates sensory analysis and sensory pleasure. The ability to discern tastes is a cultural imperative.(…) There were six different juices lined up at the head table, all covered with tinfoil. This would be blind juice tasting. The children were excited: would they be able to guess?... (…) I had been entranced not only by the fact that such a tasting existed in the first place, but also by the clever way it mimicked a traditional wine tasting and the extent to which it engaged both the children and the adults" (Trubek, 2008; pp. 45-47).

In France, policymakers and supply-chain operators have a clear perception of the centrality of educating young consumers to develop their sense of discernment about taste. This educational effort is not addressed merely to tourists: the education of rural communities to recognize the senses of sight, smell, taste, touch, and hearing is a crucial issue in avoiding the loss of tradition and for achieving an effective promotion of countryside capital.

Furthermore, trying out sensory analysis education enhances other kinds of experience, as shown by the example of the 'Festival Olioliva' in Imperia (Italy). The 'Olioliva' event is defined by its organizing committee as "a tasty stroll through the centre of Imperia discovering the historic, cultural and food values of olive oil, of bread and the typical produce of the Ligurian Riviera" (retrieved November 10, 2009, from www.olioliva.tv). It is a Festival organised by the Imperia Chamber of Commerce, the Imperia City Council and the National Association of Olive Oil Cities. The traditional products of Liguria are shown along the streets and at the same time it is possible to attend meetings with olive oil experts and didactic sensory analysis workshops and tasting workshop. Specialized tour operators organize city tours and countryside visits to discover the landscape and handicraft activities that are at the origin of the peculiar aromas previously experienced downtown. Moreover, during an entire week-end tourists can attend cooking lessons given by local chefs and visit ancient oil mills and the famous olive oil museum. In this case, the comparison of consumer (tourist) tests with the sensory profile elicited with a trained panel could be the key element in increasing the arousal of interest in the particularities of a place. In fact, by using an appropriate language the objective characteristics of a product can be described and, consequently, the consumer can tune into those elements representing the traditional qualities of the product.

Differences among local olives (or olive oils) and other products can be compared, and a wonderfully entertaining experience leads to the discovery of the natural, human and traditional factors that create such distinctiveness. In this way the meaning of *terroir* product clearly becomes a *cultural marker* of a territory.

The escapist experience regards the direct participation and immersion of tourists in events: some examples in rural tourism could be 'pick your own grape and make your wine' or 'milk a cow and drink your milk' often featured in the US.

The Castello di Verrazzano Winery in Tuscany (Chianti) organizes during harvest time tours for primary school children, offering them an *escapist experience*: children spend a morning in the vineyard with farmers, and while they are learning about vines and grapes, they collect the very grapes they will use for producing their own wine. Then they go into the winery, where the winemaker explains them how to make wine, and they crush grapes by trampling them barefoot. Next, the must is collected and each pupil gets his own bottle of grape juice together with some information about how to preserve it. Many schools have used this tool for educating pupils about rural life; however, they could achieve their educational aims even more efficiently if this escapist experience were enriched with proper sensorial analytical tools, as happens in other European countries.

This is the case with the 'Active kids get cooking' program (retrieved November 10, 2009, from www.activekidsget-cooking.org.uk), which takes place in the UK. This program aims to educate pupils to acquire a healthier life style through cooking and tasting lessons. For instance, in "the biscuit test challenger" pupils are asked to taste the difference between fat and non fat biscuits and compare nutritional labels. Other programs teach children to grow and cook their own vegetables, or educate them to eat more vegetables during the week, explaining sensory peculiarities and differences. These programs combine escapist experience – the experience of growing vegetables, visiting farms and cooking at school – with sensory education to achieve effective results.

Finally, the *esthetic experience*: an initiative like smelling the connections between wine varietals and garden flowers can fulfill a desire, and the consequent impression can be a long-lasting memory which improves tourist loyalty to a certain place when the smell is evoked again. In fact, as underlined by Distel et al. (1999), odorants that have acquired a meaning - particularly of an emotional or motivational nature - should be perceived and attended to more readily than stimuli of little relevance, and this should result in a stronger subjective perception of stimulus strength.

7 Conclusions and recommendations

In conclusion, the intensification of competition in the rural tourism business is suggesting that firms adopt a differentiation strategy. Differentiation can help rural tourism firms in:

- repositioning their offer: through differentiation companies can address their product to a market niche, as in the case of gourmet food;
- gaining a competitive advantage: the inimitability of some product attributes makes a product unique. Uniqueness and resource inimitability can help companies in achieving a sustainable competitive advantage;
- being outstanding among competitors and perceived as unique by customers (who will also be willing to pay a premium price).

The basic principles of differentiation described above can be combined with the potential offered by sensory analysis.

Why use sensory analysis? This can be conceived as a catalyst for enhancing the effects of a differentiation strategy. By offering objective evaluation standards, sensory analysis can help customers in achieving critical understanding and awareness: in this perspective sensory analysis becomes a way of acquiring customers' loyalty.

Sensory analysis allows consumers/tourists to:

1. establish among the many products/services offered which one is preferred and why;
2. use a common and objectively recognizable language for evaluating products/services and make comparisons among them;
3. give persistent reminders through memorabilia;
4. discover typical and unique characteristics of a certain product/service.

In this perspective sensory analysis can be an effective tool for designing memorable experiences; starting from the *educational* experience, sensory analysis can boost the *esthetic*, *entertainment* and *escapist* ones. Furthermore, sensory analysis can emphasize the cultural marker role of a typical product: a tourist who visits a place will have stronger memories of his impressions (and experience), if he has been educated to recognize the uniqueness of organoleptic cues.

Using sensory analysis tools in business strategy can help in succeeding in the Risk™ game of competition in the agrifood business. Recommendations for firms that want to use sensory analysis as strategic leverage could be the following:

1. firms should combine their offer of goods and services in accordance with consumer experiences; this requires a change in management orientation which should focus not only on product features but also on consumer experience;
2. a professional approach to sensory analysis should be adopted, in order to fully exploit its potential; sensory analysis is a science, and needs specific tools and training to be effectively implemented;
3. firms could use sensory analysis to prove an identity card (sensory profile) of their products in order to emphasize their features and uniqueness;
4. firms should remember that the consumer's experience can be extended over time and that memorabilia could be helpful. An example is provided by wine-tasting games, widespread in the US, which use a sensory profile for evaluating product characteristics. Organizing wine-tasting events at home enhances word-of-mouth communication.

We hope that further research will be carried out. In particular, the interaction between sensory analysis tools and managerial models are worth studying. Interesting output could come from studying the effects of sensory analysis implementation on the overall perceived quality of a service. Longitudinal studies should also be carried out on small groups of consumers/visitors in order to evaluate the effectiveness of sensory analysis tools over time.

References

Arfini, F. (2005). Segni di qualità dei prodotti agro-alimentari come motore per lo sviluppo rurale. *Agriregionieuropa 1*(3).

Aurier, P., Fort, F., and Sirieix, L. (2005). *Anthropology of food* [Online publication], May 4, 2005, retrieved March 10, 2010, from http://aof.revues.org/ index187.html.

Barney, J.B. (2001a). Is the resource-based "view" a useful perspective for strategic management research? Yes. *Academy of Management Review 26*(1), 41-56.

Barney, J.B. (2001b). Resource-based theories of competitive advantage: A ten-year retrospective on the resource-based view. *Journal of Management 27*, 643-650.

Bessière, J. (1998). Local development and heritage: traditional food and cuisine as tourist attractions in rural areas. *Sociologia Ruralis 38, 21-34.*

Briedenhann, J., and Wickens, E. (2004). Tourism routes as a tool for the economic development of rural areas—vibrant hope or impossible dream? *Tourism Management 25*(1), 71-79.

Carbone A., Gaito M., and Senni, S. (2005). *Quale mercato per i prodotti dell'agricoltura sociale?* Associazione Italiana per l'Agricoltura Biologica, Roma, Italy.

Distel, H., Ayabe-Kanamura, S., Martínez-Gómez, M., et al. (1999). Perception of everyday odors: Correlation between intensity, familiarity and strength of hedonic judgement. *Chemical Senses 24*, 191-199.

Drake, M.A. (2007). Sensory analysis of dairy foods. Invited Review: *Journal of Dairy Science 90*, 4925-4937.

Ecorys (2009). *Study on the Competitiveness of the EU tourism industry - with specific focus on the accommodation and tour operator and travel agent industries*, Framework Contract of European Commission, Sectoral Competitiveness Studies – ENTR/06/054. ECORYS SCS Group, Rotterdam.

Ecotur (2008). Sesto rapporto Ecotur sul Turismo Natura. A cura dell' Osservatorio permanente sul Turismo Natura. Bologna, Italy.

Esposti, R. (2006). Agriturismo al bivio. *Agriregionieuropa 2*(5), retrieved March 10, 2010 from http://www.agriregioni europa.univpm.it/dettart.php?id_ articolo=116

Fiore, A.M., Niehm, L., Oh, H., et al. (2007). Experience Economy Strategies: Adding Value to Small Rural Businesses. *Journal of Extension 45*(2), 1-13.

Fleischer, A., and Pizam, A. (1997). Rural tourism in Israel. *Tourism Management 18*, 367-372.

Gannon, A. (1994). Rural tourism as a factor in rural community economic development for economies in transition. *Journal of Sustainable Tourism 2*, 51-60.

Garrod, B., Wornell, R., and Youell, R.(2006). Re-conceptualising rural resources as countryside capital: The case of rural tourism. *Journal of rural studies* 22(1), 117-128.

Gatti, S., and Incerti, F. (1998). The Wine Routes as an instrument for the valorisation of typical products and rural areas, typical and traditional products: rural effects and agro-industrial problems. 52nd Seminar of the European Association of Agricultural Economists, Parma, Italy, June 19-21, 1997, Parma, Italy.

Greffe, X. (1994). Is rural tourism a lever for economic and social development? *Journal of Sustainable Tourism 2*(1/2), 23-40.

Hall, C.M., and Mitchell, R. (2001). Wine and food tourism, In Douglas, N., Douglas, N., and Derrett, R. (Eds.), *Special Interest Tourism*. Milton, Australia.

Hall, C.M., Sharples, L., Mitchell, R., et al. (Eds.) (2003). *Food tourism around the world. Development, management, and markets*. Oxford.

Hummelbrunner, R., and Miglbauer, E. (1994). Tourism promotion and potential in peripheral areas: The Austrian case. *Journal of Sustainable Tourism 2*, 41-50.
Kim, W.C., and Mauborgne, R.A. (2000). Knowing a winning business idea when you see one. *Harvard Business Review* (9-10), 129-137.
Kim, W.C., and Mauborgne, R.A. (2004). *Blue Ocean Strategy: How to Create Uncontested Market Space and Make the Competition Irrelevant*. Boston.
Kim, W.C., and Mauborgne, R.A. (2005). Value innovation: a leap into the blue ocean. *Journal of Business Strategy 26*(4), 22-28.
Kim, W.C., and Mauborgne, R.A. (2007). Blue Ocean Strategy. *Leadership Excellence* [serial online] *24*(9), 20.
Lanza, C.M., Mazzaglia, A., and Pagliarini, E. (2009). *Sensory and physicochemical characterization of different Mediterranean Malvasia wines*, III International Symposium "Malvasías", La Palma, Canary Islands, Spain, May 27-31, 2009.
Lawless, H.T., and Heymann, H. (Eds.) (1999). *Sensory evaluation of food: principles and practices*. New York.
Middleton, V.T.C., and Hawkins, R. (1998). *Sustainable tourism. A marketing perspective*. Oxford.
Moskowitz, H.R., Beckley, J.H., and Resurreccion, A.V.A. (Eds.) (2006). *Sensory and consumer research in food product design and development*. Ames.
OECD (1994). *Tourism policy and international tourism in OECD countries*. Paris.
Page, S.J., and Getz, D. (1997). *The business of rural tourism: International perspectives*. London.
Pine II B.J., and Gilmore J.H. (1998). Welcome to the experience economy. *Harvard Business Review 76*(4), 97-105.
Porter, M. (1985). *Competitive Advantage: creating and sustaining superior Performance*. New York.
Porter, M. (1998). *On Competition*. Boston.
Porter, M. (2008). The five competitive forces that shape strategy. *Harvard Business Review* 86(1), 78-93.
Reichel, A., Lowengart, O., and Milman, A. (2000). Rural tourism in Israel: service quality and orientation. *Tourism Management 21,* 451-459.
Sharpley, R., and Sharpley, J. (1997). *Rural tourism: An introduction*. London.
Sims, R. (2009). Food, place and authenticity: local food and the sustainable tourism experience. *Journal of Sustainable Tourism 17*(3), 321-336.
Somoza, A. (2007). *Manuale del turismo 2007. Dati, informazioni, indirizzi per operatori e viaggiatori*. Milano.
Trubek, A. (2008). *The taste of place: a cultural journey into Terroir*. Berkeley.
Walsh, K. (1999). Marketing and public sector management. *European Journal of Marketing 28*(3), 63-71.
Wernerfelt, B. (1984). A resource-based view of the firm. *Strategic Management Journal 5*(2), 171-180.
Williams, A., and Shaw, G. (1998). *Tourism and economic development: European experiences* (3rd ed.). Chichester.
WTO (2003). *Pathway to sustainable development*. Seminar on rural tourism, June 6-7, 2003, Kielce, Poland.

Sensory marketing research - Identification of the ideal sensory product profile

Andreas Scharf

University of Applied Sciences of Nordhausen
scharf@fh-nordhausen.de

Abstract

International suppliers of food and beverages have to answer the question in which way their products need to be adjusted to the sensory preferences of their target groups in different countries. This question also plays an important role in tourism: Every year millions of European tourists visit other European countries, and the foreigners often differ in their food preferences from the resident population. In this context, a Descriptive Panel is needed to gather all perceivable sensory product attributes and their intensities. Furthermore, an affective consumer test has to be conducted in each country in order to obtain starting points for necessary product differentiations. Only the combination of the two data sets using multivariate calibration finally provides the necessary information. For each country market those sensory attributes can be identified that influence the acceptability judgments of consumers the most. Maximizing, minimizing and optimizing these characteristics will eventually lead to an "ideal" sensory product profile for which the consumer acceptance can be estimated. This approach is illustrated by the example of optimization of frozen croquettes for the German and Italian market

Zusammenfassung

Anbieter von Nahrungs- und Genussmitteln, die international tätig sind, müssen grundsätzlich die Frage beantworten, in welcher Weise ihre Produkte an die sensorischen Präferenzen ihrer Zielgruppen in unterschiedlichen Ländern angepasst werden müssen. Diese Frage spielt auch im Tourismus eine wichtige Rolle: Jedes Jahr besuchen Millionen von Touristen andere europäische Länder, und die Ausländer unterscheiden sich häufig in ihren Nahrungsmittel-Präferenzen von der einheimischen Bevölkerung. In diesem Zusammenhang benötigt man ein Deskriptives Panel, um alle wahrnehmbaren sensorischen Produkteigenschaften und deren Intensitäten zu erfassen. Zur Gewinnung von Ansatzpunkten für notwendige Produktdifferenzierungen muss außerdem in jedem Land ein affektiver Konsumententest durchgeführt werden. Erst die Verknüpfung beider Datensätze mittels multivariater Kalibrierung liefert schließlich die erforderlichen Informationen. Für jeden Ländermarkt lassen sich diejenigen sensorischen Eigenschaften identifizieren, welche die Akzeptanzurteile der Konsumenten besonders stark beeinflussen. Die Maximierung, Minimierung bzw. Optimierung dieser Eigenschaften führt schließlich zu einem „idealen" sensorischen Produktprofil, für welches die Akzeptanz der Konsumenten geschätzt werden kann. Diese Vorgehensweise wird am Beispiel der Optimierung von Tiefkühl-Kroketten für den deutschen und italienischen Markt veranschaulicht.

K.L. Sidali et al. (eds.), *Food, Agri-Culture and Tourism: Linking local gastronomy and rural tourism: interdisciplinary perspectives*, DOI 10.1007/978-3-642-11361-1_12,

I fornitori internazionali di generi alimentari e bevande si trovano a dover rispondere alla domanda in che modo i loro prodotti necessitino di un adeguamento alle preferenze del gusto dei loro clienti in diverse nazioni. Ciò è di fondamentale importanza anche nel campo del turismo: ogni anno infatti milioni di turisti visitano altri paesi e, spesso, i loro gusti differiscono notevolmente da quelli della popolazione autoctona. In questo contesto appare particolarmente utile l'utilizzo di un panel descrittivo dei prodotti alimentari così come recepiti dagli utenti, il quale presenti tutte le caratteristiche sensoriali dei prodotti e la loro intensità. Oltre a questo strumento sarebbe auspicabile l'utilizzo di un test affettivo del consumatore propedeutico per una successiva differenziazione del prodotto. Quest'ultima può avvenire solo se si dispone delle necessarie informazioni che, a loro volta, possono essere ottenute unicamente attraverso l'abbinamento dei due data sets descritti in precedenza. Tale abbinamento si ottiene attraverso strumenti di calibrazione multivariata. Per ogni nazione si possono evidenziare nei prodotti di mercato quelle preferenze sensoriali che orientano maggiormente le scelte dei consumatori . Manipolando tali set di attributi tramite l'aggiunta o la detrazione di alcune caratteristiche sensoriali, si raggiungerà quel profilo sensoriale per così dire "ideale", da cui si può capire il gradimento del consumatore. Questo approccio metodologico viene qui di seguito illustrato tramite l'esempio di ottimizzazione delle crocchette surgelate in vendita nel mercato tedesco e in quello italiano.

Keywords: Affective consumer test; Descriptive analysis; Multivariate calibration

1 Introduction

Today, many suppliers of food and beverages act internationally. As a result they have to ask themselves whether they want to provide standardised products or offer differentiated ones, and thereby account for consumers of different countries and their different preferences. When it comes to international marketing the rule applies (Levitt, 1983): as much standardisation as possible, as much differentiation as necessary. Sensory marketing research leads to the right decision of sensory product standardisation versus differentiation. It is the systematic investigation of the functional relationship between sensory perceptions (e.g., crunchiness, sweetness) and corresponding consumer reactions (e.g., preference, acceptance).

In order to decide the degree of standardization and differentiation of sensory characteristics of products to be offered in different countries, two different perspectives need to be linked: a Descriptive Panel can ideally be used to measure the product perception, but this tool should not be used to measure the acceptance (Stone and Sidel, 2004). Consumers are able to indicate product acceptance, but they should not be used to measure the objective product perception (Meilgaard et al., 2006). In other words: while consumers from different national markets evaluate the products affectively, sensorial trained individuals, that is, a Descriptive Panel, describe their perception of the sensory characteristics of these products.

This approach is based on the assumption that consumers are well able to distinguish different sensory products in terms of their benefits. As a result, consumers perceive the sensory attributes of products and make their affective judgments based on their perceived embossing of these characteristics (Greenhoff and MacFie, 1994; Lawless and Heyman, 1999). Due to a low cognitive control over the recording and processing of sensory product information, consumers cannot describe their sensory perception reliably. Therefore, the identification and quantification of all relevant sensory product characteristics is done by descriptive analysis (Scharf, 2000). If detailed pieces of information exist as to how the sensory attributes of the test products differ, the multivariate linkage with affective data allows for identification of sensory attributes and their intensity which are responsible for the preference or rejection of certain products. Knowledge of these connections is a requirement for the systematic optimization of sensory product characteristics and can be used in the context of international marketing. Figure 1 illustrates this at a glance:

Figure 1. Identification of the ideal product profile by using multivariate calibration

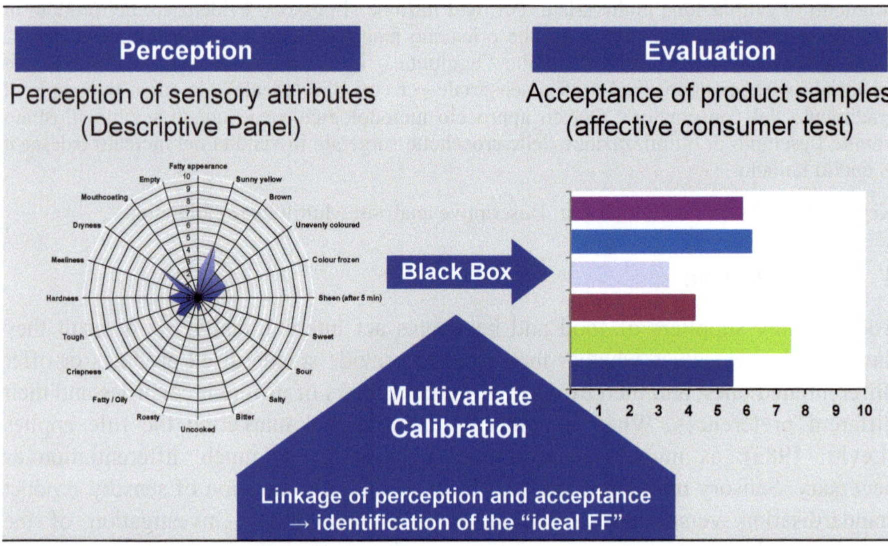

2 Linking perception data and acceptance data

2.1 Measuring perception - Descriptive analysis

Making sensory perception objective and precisely quantifiable is an important goal. The so-called descriptive methods of sensory product research make it possible to do so by identifying and quantifying product properties (Piper and Scharf, 2004). They should not be confused with other methods of sensory product research – like those used to identify sensory product differences (Bi, 2006; O'Mahony, 1986)

Every descriptive method is performed by a specially trained group of people – frequently termed a panel. So humans serve as the measurement instrument, whose objectivised perception will enable sensory product properties to be reliably identified and quantified (Scharf, 1996). Finally, the resulting sensory profile of a product allows a direct link to be established between the product's recipe and its affective evaluation by potential or actual customers (Stone and Sidel, 2004; Murray et al., 2001), making the Descriptive Panel an important instrument at the interface between marketing, market research and product development.

Today, a descriptive panel is usually made up of consumers, even though training them is initially a time-consuming and costly process. The reason for this can be seen from the fact that experts possess different perceptive structures than consumers due to their many years of experience and detailed knowledge of recipes (Stone and Sidel, 2004). Moreover, in addition to their sensory perception, experts also take into consideration their knowledge of the underlying chemical-physical attributes. Further disadvantages of using experts are their professional involvement, their so-called "corporate taste", and sometimes their lack of feeling for sensory trends.

Sensory product properties are generally described in two steps: First qualitatively, then quantitatively. Panelists are selected primarily on the basis of their sensory abilities, because after attending the training phase they have to be able to identify all sensory properties of test products and precisely verbalize them (Scharf, 1996). Other authors highlight the importance of the commitment and the motivation of the panelists for the overall success of the endeavor (Murray et al., 2001). In the second step, the participants additionally learn to precisely indicate and consistently formulate intensities. The descriptors that are used then have to be translated into the language of product developers (Civille and Lawless, 1986). Several sources provide a good overview of the descriptive methods that have been developed over the course of more than 50 years (Hootman, 1992; Powers, 1988; Scharf, 2000). Descriptive analysis supplies RandD and marketing with detailed sensory profiles visualized by so called spider plots, revealing not only the perceivable components and their intensities, but also the differences among the samples. In short Figure 2 shows the different steps of descriptive analysis.

2.2 Measuring acceptance - affective consumer test

While procedures for determining product profiles start with the human perception, the objective of the so-called affective consumer tests is the detection of the emotional appreciation of relevant subjects regarding the perceived sensory attributes of products (Meilgaard et al., 2006; Stone and Sidel, 2004). Within sensory product research an affective test, like the descriptive analysis, is a study designed as a blind test, which means the test products are marked only by a code. More detailed information exceeding the product core, such as brand or price, is therefore not available to the test subject (Scharf, 2000).

Over time, a number of different approaches for the identification of affective judgments of consumers have been developed which vary towards the goal of the investigation, duration and location of conduct, the nature of the assessment task, the scaling method, the number and presentation of the stimuli as well as the selection and size of the sample. Of fundamental importance is the classification of the affective tests regarding the assessment task and the scale in sensory preference and acceptance tests (Amerine et al., 1965; Lawless and Heymann, 1999). In an acceptance test consumers receive samples (mostly monadic or sequential monadic) and indicate, how much they like each sample (indirect comparison).

Figure 2. Overview of the typical steps of descriptive analysis

Selection of panellists "Screening"	▪ Requirement: 8-12 persons with above-average sensory skills ("Descriptive Panel"). ▪ Selection of panellists is done in a 2-step product related screening phase. ▪ Step 1: Invitation of about 200 persons to 1 hour session; realisation of several tests (see ASTM guidelines, e.g. basic taste test, odour identification test [GITU], odour recognition test [GELTU], triangle tests, verbalization test etc.) ▪ Step 2: Short 2nd testing session with the best 20-40 participants under more severe conditions; personal interview with panel leader; interview with the panel leader about motivation, investigation of verbalisation skills and personality; definite selection of 10-12 panellists.
Training of Panellists	▪ First session = prelude session (get acquainted; organisational arrangements). ▪ Development of a descriptive language (individual description of all samples, collection and discussion to find a consistent list of terms); always close communication with R&D. ▪ Selection of reference-stimuli and -intensities; isi philosophy is to use references in the shape of the product, if possible. ▪ Stepwise training of scale usage; isi uses unstructured 10 cm line scales with only 2 verbal anchors (0= not detectable; 10= very intensive).
Test Measurements	▪ All samples are profiled under testing conditions (red light, usage of booths etc.). ▪ Statistical Analysis of panel performance, using sophisticated statistical software like Senstools, SPSS etc. (Cluster Analysis, Analysis of Variance, Procrustes Analysis). ▪ Individual and collective feedback is given by panel leader. ▪ Further training sessions, if required.
Descriptive Profiling	▪ Balanced presentation order. ▪ All samples are profiled with replications to check consistency of data. ▪ Detailed data analysis and visualisation of findings, professional research report with management summary, if required.

This results in the degree of acceptance on a hedonic scale or a "just about right" scale. In a preference test the samples are jointly presented and evaluated (direct comparison). Consumers rank the different products according to their overall preference or according to single sensory dimensions (e.g., appearance, odor). Acceptance tests deliver data that can be extensively quantitatively analysed, but preference tests are more intuitive and easier to handle, especially for particular target groups (e.g., children) (Stone and Sidel, 2004).

The scale with the greatest empirical relevance is the nine-point hedonic scale. This scale is bipolar, verbally anchored, and includes a neutral response option ("neither like nor dislike"). In addition, the nine-point hedonic scale is balanced, meaning the number of positive and negative response options is even. The word anchors for the nine stages are based on a single word stem, and the adjectives used for gradation are identical in the positive and negative area. The nine-point hedonic scale is based on the assumption that consumers can transform their emotional appreciation for the sensory attributes of a product in discrete, positively and negatively verbalized categories of a scale (Meiselman, 1988). Figure 3 illustrates the 9-point hedonic scale.

Figure 3. 9-point hedonic scale

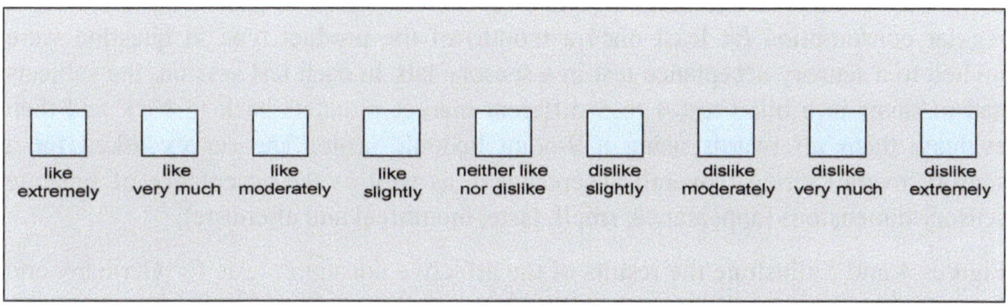

2.3 Linking both data sets - Multivariate calibration

Acceptance tests answer several questions concerning the liking/disliking of products or prototypes; and give detailed insight into the sensory perception of particular products by experts using a common language. For product development purposes, it is critical to know which sensory perceptions are responsible for consumers´ judgements.

To establish a functional relationship between the results of a Descriptive Panel and consumer judgements one can use Partial-Least Squares Regression (PLS-R). PLS-R finds a linear regression model by projecting the observed variables X (descriptive data) and the predicted variables Y (affective data) into a common space. In addition it is possible to predict acceptance scores for new products without conducting an affective test, solely by application of the functional relationship (Martens and Naes, 1991).

Multivariate calibration of descriptive and affective data answers the following questions of food and beverage producing companies, which sell their products in different countries: Which sensory dimensions of a new or modified product have a positive preference or acceptance effect ("positive drivers") in which country? Which sensory dimensions effect the consumers' affective judgements negatively (e.g., off-flavours)? Which intensity of a sensory dimension (e.g., saltiness, sweetness) is optimal from a consumer´s perspective? Do similar preference structures exist in different countries, so that consumers with similar sensory likes and dislikes can be merged to reasonable segments?

3 Identification of consumer acceptance drivers of frozen croquettes – an empirical approach

3.1 Consumer acceptance of frozen croquettes market products in Germany and Italy

The goal of the following empirical study was to analyze whether differences between German and Italian consumers exist regarding their preferences for the sensory attributes of frozen croquettes which were prepared in an oven. For reasons of confidentiality the data presented here is masked.

In various German and Italian cities 120 consumers, who had previously indicated a regular consumption (at least once a month) of the product type in question were invited to a sensory acceptance test in a sensory lab. In each test session, the subjects had to savor in a blind test 4 to 5 different market products back to back and then evaluate them afterwards using a 9-point hedonic scale. The survey asked for a sensory overall rating ("overall acceptance"), as well as the acceptance of separate sensory dimensions (appearance, smell, taste, mouthfeel and aftertaste).

Figures 4 and 5 illustrate the results of the affective consumer tests for Germany and Italy. It should be noted that only a few items that were tested in the two countries were sensorially identical. In Germany, 4 market products of the considered company (light blue) were evaluated and 9 competing products (dark blue). Market product A got the best acceptance score (mean score: 7.7) as well as the product of competitor 1. Some of the German market products are evaluated relatively poor by consumers, especially the competing products 10 and 11. In Italy, consumers assess the competing product 1 (mean score: 7.8) significantly better than the best product A of the considered company (mean score: 7.6). Overall, the differences in acceptance are considerably lower in Italy than in Germany, which is due to minor sensory differences between the market products.

Figure 4. Overall acceptance of 15 frozen croquettes market products in Germany (sensory lab test; mean scores; n = 360)

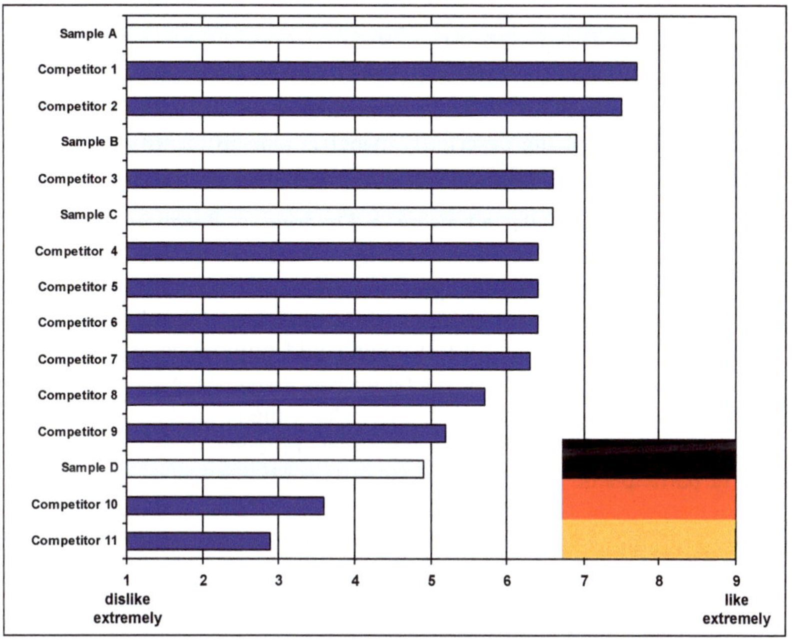

Figure 5. Overall acceptance of 13 frozen croquettes market products in Italy (sensory lab test; mean scores; n = 360)

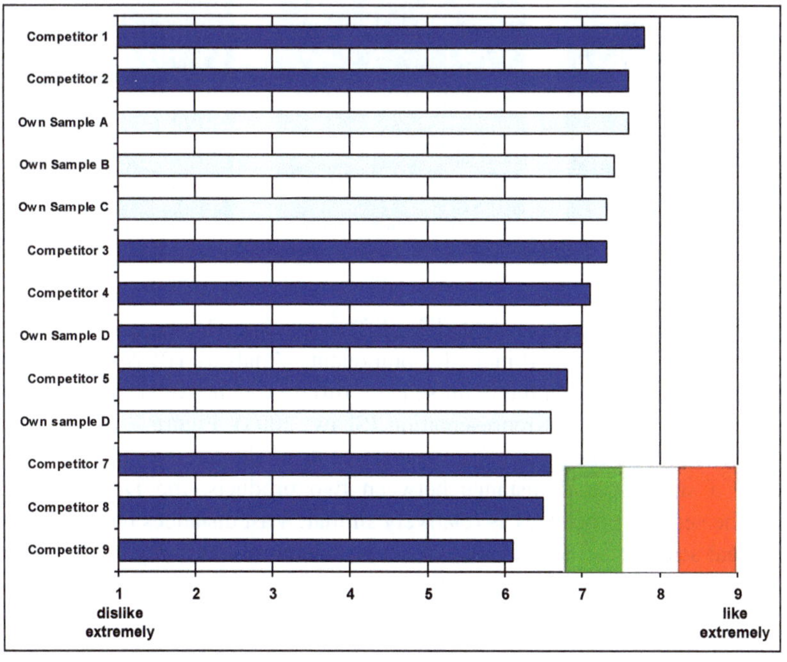

3.2 Perception of frozen croquettes market products by a descriptive panel

In international studies, the affective data must always be collected in the country markets of interest, in this case in Germany and Italy. In comparison, the location of the Descriptive Panel or the nationality of sensory experts can be chosen freely because the goal of the analysis is the objectification of human perception and not the assessment of the products in question. The quality of the perception data thus depends only on how well the descriptive panel is able to identify all the sensory attributes of the test products and state their perceived intensity on a scale.

The perception data was collected using a Descriptive Panel in Germany. The experienced panel consisted of 12 sensorially trained consumers who have been involved for several years with the profiling of frozen potato products. For the sensory description of the market products, the expert group developed a list of so-called descriptors, through which all test items could fully be described. The list of terms used is shown in Figure 6. Consideration was given to the sensory dimensions "appearance", "taste" and "texture". In order to check the reliability of the perception data in a better way, all the German and Italian test products were analyzed twice by each of the 12 participants.

Figure 6. The Descriptive Panel used 18 attributes to measure the perception of the German and Italian croquettes samples

Appearance	Taste	Texture
fatty appearance	sweet	crispness
sunny yellow	sour	tough
brown	salty	hardness
unevenly coloured	bitter	mealiness
	uncooked	dryness
	roasty	mouth-coating
	fatty / oily	empty

The descriptive data can be evaluated in different ways. Compared to mean score tables and spider webs the Principal Components Analysis (PCA) is particularly advantageous as it visualises the main sensory differences between the products in a low-key dimensioned spatial representation (Shaw, 2003). Figure 7 illustrates as an example the results of the perception measurement for the German market products. In case of a small spatial distance between two products, the Descriptive Panel perceives the sensory properties as relatively similar. Vast distances between products indicate great sensory differences. A product is characterized mainly by those sensory attributes which have a similar spatial orientation. For example, all products which are particularly "sunny yellow" are located on the right side of the graphics. The representation locates market product E and competing product 8 at the bottom, because they both are perceived as "fatty/oily", "empty", "sour" and "crispy".

Figure 7. Perception of the German croquettes samples visualized by PCA (Explanation of 78 percent of the differences with two dimensions)

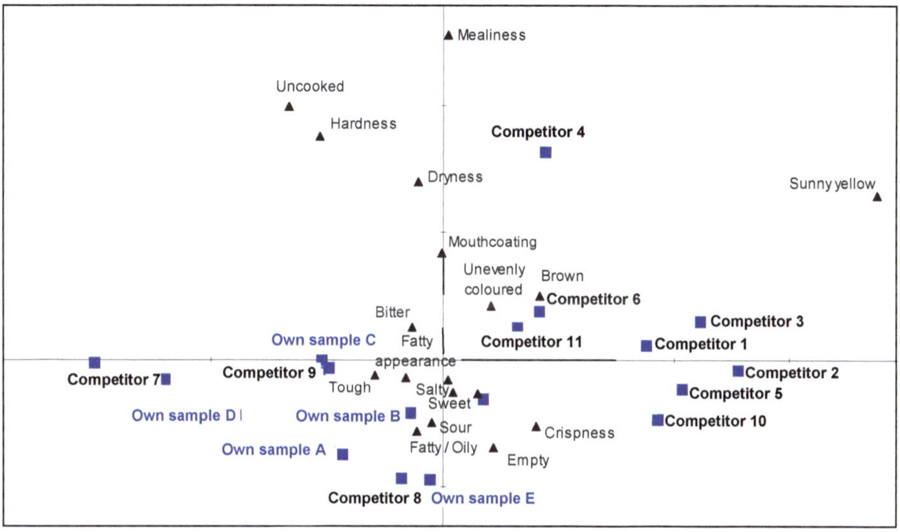

3.3 Identification of the "ideal" croquettes product for the German and Italian market

The linkage of descriptive and affective data is carried out in two steps. First the researcher tries to identify those descriptors which impact the consumer acceptance significantly. Secondly the "ideal" intensities of these "relevant" descriptors within the frame of reference must be identified. Partial-Least-Squares Regression was used to establish the functional relationship between the descriptive data (X) and the affective data (Y).

Figure 8 visualizes the results of the first analysis step for the German and the Italian market. With regard to those sensory attributes, which are particularly important for the sensory assessment of the market products by the consumers, both countries show considerable differences. For example, regarding the appearance of frozen croquettes German users pay attention to a "sunny yellow" colour, while for Italians it is important that a product is not "unevenly coloured". There are also similarities in terms of the preference drivers because in both countries "crispness" is one of the most important descriptors for the acceptance.

Figure 8. Sensory attributes which influence the acceptance of croquettes products significantly

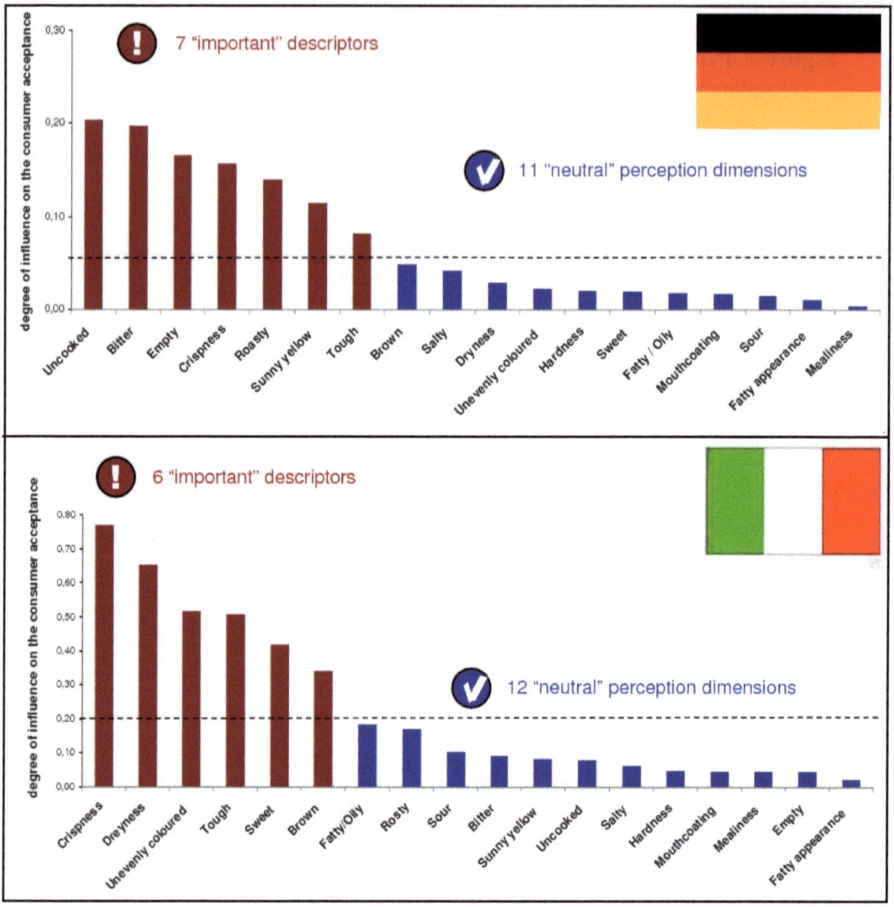

To answer the question about the "ideal" product for a particular country, the direction of optimization for each identifiable preference driver needs to be taken into consideration. For positive preference drivers it is valid, that the higher the attribute's intensity the higher is the consumer acceptance, the opposite functional relationship is valid for negative preference drivers. But there are also existing sensory attributes which show an optimum level. Here, consumer acceptance increases with increasing intensity. At a certain point the optimum is reached and higher intensities than the optimum result in decreasing acceptance. Figure 9 illustrates the recommendations for product development to produce the "ideal" frozen croquettes for the German market. For example, it is important to maximize the crispness, to minimize the bitter taste and to optimize the toughness of the product.

Figure 9. The important attributes affect the consumers' acceptance differently

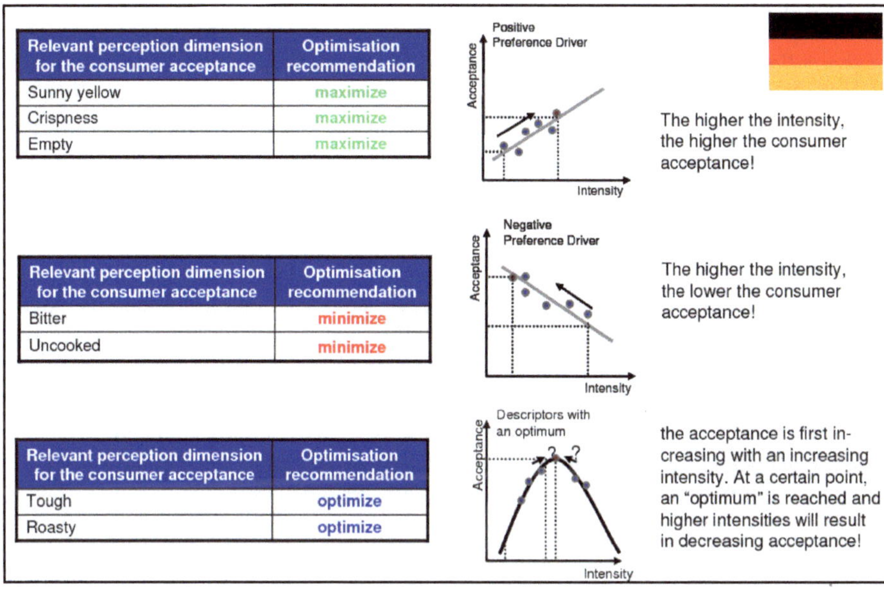

Finally the researcher can predict the acceptance scores for the optimized prototypes solely by application of the functional relationship between perception data (X) and affective data (Y). Figure 10 visualizes the difference between the current best product and the sensorial optimized product for the German and the Italian market. By means of multivariate calibration an acceptance value of 8.2 can be predicted for the optimized German product. The optimized Italian product will in all likelihood reach a value of 8.6.

Figure 10. Optimization of the important sensory attributes results in an increase of consumer acceptance

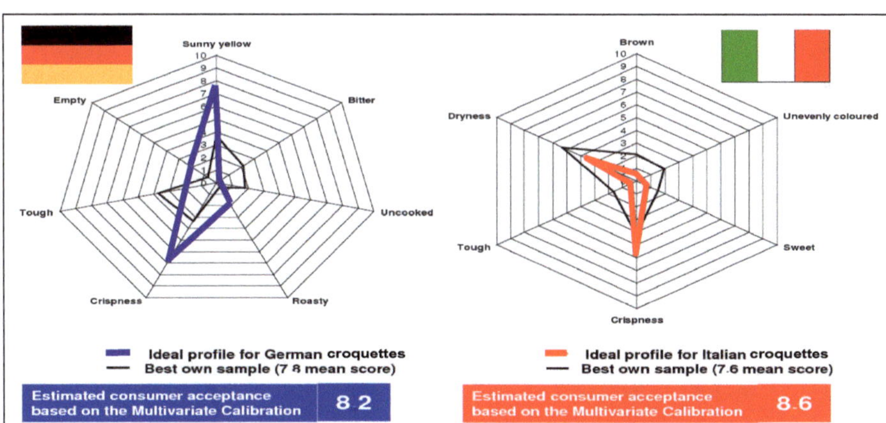

4 Conclusion

For the collection of the sensory perception of products it is common today to use a Descriptive Panel and capture the affective judgments of relevant target groups by means of acceptance tests in a central location or a sensory lab. However, only the systematic linkage of the two data sets using multivariate statistical analysis provides the necessary information for sensorial standardization or differentiation procedures in international marketing. Meanwhile, efficient software solutions are available (eg. UNSCRAMBLER, XLSTAT), which greatly facilitate this task.

The procedures outlined above allow for each country market to identify those sensory attributes that influence the acceptability judgments of the consumers the most. Maximizing, minimizing and optimizing these characteristics finally lead in for each country to an "ideal" sensory product profile, for which the consumer acceptance can be estimated. Companies that apply this systematic approach need to spend substantial financial resources for market and sensory research, but they also hold the crucial key to international market cultivation of food and beverages.

References

Amerine, M.A., Pangborn, R.M., and Roessler, E.B. (1965). *Principles of sensory evaluation of Food.* NewYork: Academic Press.

Bi, J. (2006). *Sensory Discrimination Tests and Measurements: Statistical Principles, Procedures and Tables*. Ames: Blackwell Publishing.

Civille, G.V., and Lawless, H. (1986). The importance of language in describing perceptions. *Journal of Sensory Studies, 1*, 217-236.

Greenhoff, K., and MacFie, H.J.H. (1994). Preference mapping in practice. In MacFie, H.J.H., and Thomson (Eds.), *Measurement of food preferences* (pp. 137-166). London: Black Academic and Professional.

Hootman, R.C. (1992). *Manual on descriptive analysis testing for sensory evaluation*, ASTM manual series, MNL 13, Philadelphia: Association of Sensory Testing and Materials.

Lawless, H., and Heymann, H. (1999). *Sensory Evaluation of Food. Principles and Practices.* New York: Chapman and Hall.

Levitt, T. (1983). The Globalization of Markets. *Harvard Business Review, 61*(5), 87-91.

Martens, H., and Naes, T. (1991). *Multivariate Calibration.* Hoboken: Wiley.

Meilgaard, M., Civille, G.V., and Carr, B.T. (2006). *Sensory Evaluation Techniques.* Boca Raton: CRC Press.

Meiselman, H.L. (1988). Consumer studies of food habits. In Piggott, J.R. (Ed.), *Sensory analysis of foods* (pp. 267-334). London: Elsevier Applied Science.

Murray, J.M., Delahunty, C.M., and Baxter, I.A. (2001). Descriptive Analysis: past, present and future. *Food Research International, 34*, 461-471.

O'Mahony, M. (1986). *Sensory evaluation of food. Statistical methods and procedures.* New York: Marcel Decker.

Piper, D., and Scharf, A. (2004). Descriptive analysis – state of the art and recent developments. In Scharf, A. (Ed.), *Series Sensory Analysis, 1*, Göttingen.

Powers, J.J. (1988). Current practices and application of descriptive methods. In Pigott, J.R. (Ed.), *Sensory analysis of foods* (pp. 187-266). London: Elsevier Applied Science.

Scharf, A. (1996). Verfahren der sensorischen Produktforschung. In Knoblich, H., Scharf, A. and Schubert, B. (Eds.), *Geschmacksforschung. Marketing und Sensorik für Nahrungs- und Genußmittel* (pp. 131-183). München: Oldenbourg.

Scharf, A. (2000). *Sensorische Produktforschung im Innovationsprozess.* Stuttgart: Schäffer-Poeschel.

Shaw, P. (2003). *Multivariate Statistics for the Environmental Sciences.* London: Arnold Publishers.

Stone, H., and Sidel, J.L. (2004). *Sensory Evaluation Practices.* San Diego: Academic Press.

Conclusions

Katia L. Sidali

University of Göttingen
Department of Agricultural Economics and Rural Development
Katia-Laura.Sidali@agr.uni-goettingen.de

The preceding contributions have revealed different "realities of farm and rural tourism" as well as multiple possibilities of its interweaving with other forms of tourism such as culinary tourism as well as other aspects of experiencing "nature" and "agrarian life" such as on educational farms. This final chapter explores the extent to which these themes can be placed in a contextual landscape to enhance further empirical research. The conceptual framework that we are going to introduce focuses on individuals' consumption patterns and shows how these shape the nature of farm tourism. At first glance the relationship between farmers and tourists, which is the core of farm tourism, appears to be based largely on contrasts. For instance, Getz, Carlsen and Morrison (2004) stress that farming is "supply-driven", whereas tourism is "market-led". Furthermore, farmers travel the least, rural tourists on the contrary travel very frequently. Henceforth, market and consumer orientation is a difficult goal to reach.

According to many scholars (Hill and Busby, 2002; Roberts and Hall, 2004), in order to understand this relationship, research should concentrate on the sense of "disconnection" from the natural world and agriculture that many people feel. As Adams (2008) observes, "the more humans become removed from nature, the more they try to reconnect to it". This "reconnection" process passes through individuals' interpretation of nature which is, first of all, a dynamic process. Knudsen and Greer (2008) point out that, before the romantic movement of the second half of the 18th century, mainstream Western aesthetics was shaped by the classical concept of beauty. As a consequence, nature was considered "too imperfect" because it was "lacking in symmetry" (Kwa, 2005). Common tourist destinations of the 21st century, such as the seaside and wooded areas, were considered wild and somewhat frightening, and it is only since industrialization that they have become interesting (Romeiß-Stracke, 1998). Furthermore, people's perception is influenced by the image of nature common in their country. Thus, for example, Germans are much fonder of rambling than Italians (ibid.).

A number of scholars have tried to identify common interpretations of nature (see Figure 1). The mystique associated with rural areas is a recurrent topic in the literature (Hjalager, 1996; Wilson et al., 2001). Thus, nature is filtered through a feeling of nostalgia, which, according to Romeiß-Stracke (1998), is an indefinable quest for a real or imaginary place where a person feels protected. Thus, according to the literature, consuming the countryside (Roberts and Hall, 2004) should be seen as a response to the individual's need to "immerse oneself in the emotion of nostalgia" (Hjalager, 1996).

K.L. Sidali et al. (eds.), *Food, Agri-Culture and Tourism: Linking local gastronomy and rural tourism: interdisciplinary perspectives*, DOI 10.1007/978-3-642-11361-1_13,

Figure 1. The functions of farm tourism as reconnection to nature

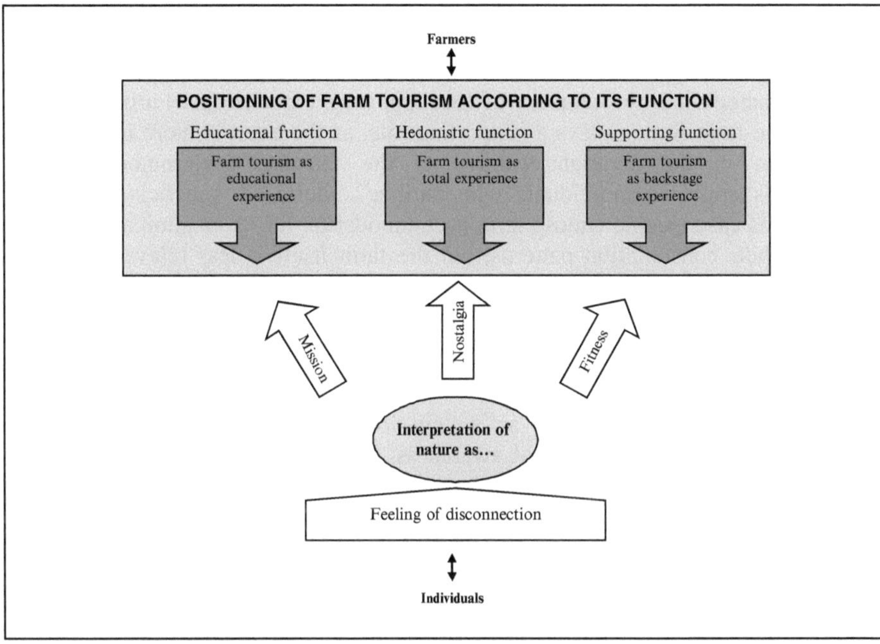

Source: own representation based on Romeiß-Stracke (1998)

Farm tourism associations and other, related businesses are conscious that this type of tourism is attractive because it provides "what urban life cannot give" (Nilsson, 2002). As a consequence, farm tourism should fulfill a hedonistic function; after all, even on a farm holiday, tourists expect not only basic requirements but also memorable experiences (Hill and Busby, 2002). However, Hjalager (1996) stresses that the "modern agricultural community is not immediately capable of providing the facilities enabling tourists to experience or re-experience even the very near past [of agrarian life]." Hence, farm tourism associations, as well as tour operators and farmers, have the essential task of clustering a variety of products and activities in order to "reinvent tradition" (ibid.) and, in this way, engrave the farming environment on the consciousness of farm guests. The latter is particularly important because, nowadays, it is not the holiday product *per se* but human recollections that guests are willing to pay for. The quality labels "Erlebnis Bauernhof" ("Adventure Farm") in Germany and "agriturismo ospitalità" ("Hospitality Farm") in Italy are indicative of this attempt to court the favor and fulfill the emotional needs of farm guests (Hill and Busby, 2002).

If, on the one hand, many people have a romantic vision of nature and want to be part of it, on the other hand, others share a more mechanistic approach and want to "utilize" nature for reasons such as its positive repercussions for physical and psychological well being. This is the case for many tourists who choose rural areas for practicing outdoor sports and other health-related activities. This perception of nature as the "fitness center of the human being" (Romeiß-Stracke, 1998) has been sensed by the farm holiday sector, where the idea of "adding healing arts (bodywork, day spas etc.) to the farm's agritourism package has been emerging" (Adams, 2008). In practice, farms with this specialization could host health-related activities, such as yoga lessons, training in the creation of home spa and cosmetic recipes and sport-themed workshops led by members of local sporting associations. Furthermore, in such cases the farm setting plays a supporting role, as the guest is more interested in relaxing than in the agrarian features of the farming environment. Similar considerations apply to individuals who "utilize" nature for practicing "extreme sports". In this case, people choose farm accommodation for its location and because it fits with their consumption patterns, but the farm itself is less relevant than the activity and sometimes almost irrelevant (cf. Roberts and Hall, 2004). In this context, whether alone or in collaboration with external businesses, farmers make their natural surroundings available to adventure seekers by providing a variety of services, such as camping facilities and short-term accommodation during extreme sports tours.

Finally, nature can be perceived as a "mission" as in the case of individuals characterized by high environmental awareness. This, according to Romeiß-Stracke (1998), originated in the ecological movement of the 1970s and 1980s. Given the high priority such guests place on nature issues, they hope to have a close relationship with the farmer. These individuals are aware of the farm. Moreover, they are interested in being connected with agrarian life through an educational approach and not through a quest for peak experiences. On the practical level, whereas the pleasure-seeker (Schulze, 1992) can be interested in experiencing outdoor fires or participating in workshops, such as picnic preparation or gift-making on the farm (cf. Adams, 2008), the education-seeking farm guest looks for other types of activities, such as the opportunity to work on the farm, to learn organic farming techniques (adults) or to take care of the animals (children).

The different interpretations of nature correspond to different needs of individuals; accordingly, farmers should try to fulfill these needs by placing their facility in a position to meet these objectives. The food component is a signifier of the agrarian culture and, at the same time, a "joyful experience" for body and soul which forms a bridge between educational (knowledge of food sources) and hedonistic (sensory experience) functions. The preceding contributions have shown manifold ways in which farmers can build up this bridge both alone or together with special interest actors such as the Slow Food movement, etc. Italian farmers, for instance, were pioneers in providing their guests with high-value oeno-gastronomical specialties which reflect deeply-rooted traditions in the rural area. In the same way, German farm operators whose farms are located in highly valuable wine regions (e.g., Saale-Unstrut) have started orienting their farms towards the regional landscape. Furthermore, the increasing success of educational farms indicates a strong desire within society to explore farm life, to improve its awareness of environmental matters (such as knowledge of food processing traditions) and to be "educated" by farmers.

In conclusion the contributions of this book have advanced knowledge about the complex interfacing between agrarian life and culture, the rural and farm tourism sector as well as culinarity issues. By reviewing both the theory and practice this book has contributed to the development of a better approach to this interdisciplinary field.

References

Adams, B. (2008). *The new agritourism. Hosting community and tourists on your farm.* Auburn, California: New World Publishing.

Getz, D., Carlsen, J., and Morrison, A. (2004). *The family business in tourism and hospitality.* Oxfordshire: CABI.

Hill, R., and Busby, G. (2002). An inspector calls: farm accommodation providers' attitudes to quality assurance schemes in the county of Devon. *International journal of tourism research, 4*(6), 459-478.

Hjalager, A. M. (1996). Agricultural diversification into tourism. *Tourism Management, 17*(2), 103-111.

Knudsen, D. C. and Greer, C. E. (2008). Heritage tourism, heritage landscapes and wilderness preservation: the case of National Park Thy. *Journal of Heritage Tourism, 3*(1), 18-35.

Kwa, C. (2005). Alexander von Humboldt's invention of the natural landscape. *The European legacy, 10*(2), 149-162.

Nilsson, A. P. (2002). Staying on farms—An Ideological Background. *Annals of Tourism Research, 29*(1), 7-24.

Roberts, L., and Hall, D. (2004). Consuming the countryside: marketing for "rural tourism". *Journal of Vacation Marketing, 10*(3), 253-263.

Romeiß-Stracke, F. (1998). Vorwärts: Zurück zur Natur? Trends im Tourismus und ihre Konsequenzen für den Urlaub auf dem Lande. In H.-G. Burger, G. Diehl and M. Packeiser (Eds.), *Der deutsche Landtourismus - Wege zu neuen Gästen* (Vol. 3). Frankfurt a. M.: DLG (Deutsche Landwirtschafts-Gesellschaft).

Schulze, G. (1992). *Die Erlebnisgesellschaft: Kultursoziologie der Gegenwart.* Frankfurt a.M.: Campus.

Wilson, S., Fesenmaier, D. R., Fesenmaier, J., and van Es, J. C. (2001). Factors for success in rural tourism development. *Journal of Travel Research, 40*(2), 132-138.

MIX
Papier aus verantwortungsvollen Quellen
Paper from responsible sources
FSC® C105338

If you have any concerns about our products,
you can contact us on
ProductSafety@springernature.com

In case Publisher is established outside the EU,
the EU authorized representative is:
Springer Nature Customer Service Center GmbH
Europaplatz 3, 69115 Heidelberg, Germany

Printed by Libri Plureos GmbH
in Hamburg, Germany